# PALAEOPATHOLOGY

PAUL A. JANSSENS

# PALAEOPATHOLOGY

Diseases and injuries of prehistoric man

5 ROYAL OPERA ARCADE
PALL MALL LONDON S.W.1

© 1970 PAUL A. JANSSENS

Translated into English by Ida Dequeecker

First published in 1970 by
JOHN BAKER (PUBLISHERS) LTD
5 Royal Opera Arcade
Pall Mall London S.W.1

SBN 212 99844 7

First Published in the U.S.A. by
HUMANITIES PRESS INC
303 Park Av South
New York, N.Y. 10010
U.S.A.

Printed in Great Britain at
THE CURWEN PRESS
Plaistow London E.13

to my wife NINI

'Speculations usually precede discovery. Suggestions and theories precede definite concepts.' ROY L. MOODIE

# Contents

| | | |
|---|---|---:|
| | FOREWORD | xi |
| 1 | DEFINITION | 1 |
| | Definition of medicine | 2 |
| 2 | INFLUENCE OF CLIMATE, ECONOMY AND ENVIRONMENT | 4 |
| | Relation with religion | 5 |
| 3 | DEFINITION OF DISEASE | 6 |
| 4 | CHRONOLOGY OF THE QUATERNARY | 9 |
| 5 | MATERIAL | 14 |
| 6 | THE IMPORTANCE OF PALAEOPATHOLOGY | 16 |
| 7 | SKELETAL REMAINS | 19 |
| 8 | TRAUMATA | 25 |
| 9 | INJURIES FROM STONE WEAPONS | 35 |
| 10 | PREHISTORIC ART | 38 |
| 11 | PREHISTORY—TRADITION AND FOLKLORE | 45 |
| 12 | DISTURBANCES IN DEVELOPMENT | 47 |
| | Metabolic disturbances | 55 |
| 13 | 'VENUS' FIGURINES | 56 |
| 14 | AVERAGE DURATION OF LIFE | 60 |
| 15 | DEFICIENCY DISEASES | 64 |
| | Rickets | 64 |
| | Scurvy | 66 |
| 16 | TUMOURS | 67 |
| 17 | REACTIONS | 71 |
| | Immunization | 71 |
| | Infection | 72 |
| 18 | ARTHRITIS | 75 |
| | Gout | 89 |
| 19 | DENTAL DISEASES | 90 |
| 20 | TUBERCULOSIS | 98 |

21 SYPHILIS                                        103
22 POLIOMYELITIS                                   111
23 OTHER SPECIFIC INFECTIONS                       114
    Plague                                         114
    Smallpox                                       114
    Leprosy                                        114
    Malaria                                        115
    Sleeping sickness                              115
    Actinomycosis                                  115
    Goundou                                        115
    Verruga Peruana                                116
    Uta                                            116
    Nigua                                          116
24 DISEASES OF THE SOFT TISSUES                    117
25 MEDICAL VIEWS ON PREHISTORIC
   REPRESENTATIONS OF HUMAN HANDS                  120
26 TREPANATION                                     125
    Sincipital-T. Bregma-Nerven                    138
27 THERAPY                                         140
   CONCLUSION                                      150
   GLOSSARY                                        151
   BIBLIOGRAPHY                                    159
   INDEX                                           167

# Plates

1   Cranial silhouette from a bell-beaker grave.
2   'Bog corpse' of the third to fifth century from Grauballe, Denmark.
3   Plaster cast of a victim of the Vesuvius eruption.
4   Flint knives used by Egyptian embalmers.
5   Goat horn, mounted in silver.
6   *Bâton de commandement* with engraved figures, made of deer horn.
7   Radiograph of a tibia.
8   Piece of bone from the burial cave at Furfooz.
9   Cremation remains from a burial mound at Hogeloon, The Netherlands.
10  Part of a reindeer skull with characteristic opening caused by a
    Lyngby axe (below).
11  Diagram of the lesions on a skull from Rhodesia.
12  Skull of a man with three impacted fractures.
13  Skeleton of a miner.
14  Four ulnae with well consolidated fractures.
15  Radiographs of four ulnae.
16  Human vertebra in which is lodged a Neolithic arrowhead.
17  *Elephas primigenius*.
18  Deer from cave of Covalanas, Spain.
19  Arrowheads of the upper Solutrean culture.
20  Arrowheads found in the Belgian Kempen, dating from the
    Neolithic period.
21  Pieces of bows.
22  Wooden shafts of arrows.
23  Pitfall, partly surrounded by a series of dots.
24  Tectiform figures with series of dots.
25  Bison wounded in the belly by a stabbing weapon.
26  The well-known hunting scene from the cave of Lascaux.
27  Drawing of a *baguette décorée*.
28  Decorated object in reindeer horn.
29  Decorated bird bone.
30  Deer and fishes engraved on a piece of bone.
31  Part of the sternum of a male skeleton found at Avennes.
32  King Siptah's mummy.
33  Femur affected by Paget's disease.

34 Bilateral osteoporosis.
35 Venus of Willendorf.
36 Venus of Lespugue.
37 Venus of El Pendo.
38 Osteoma of the rib.
39 Osteosarcoma.
40 Osteitis of the whole lower jaw, originating from a dental abscess.
41 Carcinoma of the rhino-pharynx.
42 Tuberculum of Carabelli.
43 Adult man with Aymara cranial deformation; also alveolar abscess
   of the mandible.
44 Pott's disease.
45 Pott's disease in a priest of Ammon.
46 Tibiae with syphilitic lesions; tibiae and humerus with
   identical lesions.
47 Pre-Columbian skull with syphilitic lesions.
48 'Boomerang leg' in a woman of forty from Central Australia.
49 Stele of the Priest Ruma.
50 Lesion in the leg of a boy of eight, the result of poliomyelitis
   in early childhood.
51 Negative drawing of a hand above the picture of a horse.
52 Representations of mutilated hands.
53 Representation of a mutilated hand.
54 Diagram of the most frequent mutilations at Gargas.
55 Congenital opening of the os parietale.
56 Skull with a case of dysostosis cleido-cranialis.
57 Meningocoele.
58 Tuberculosis of the cranial roof.
59 Syphilis of the cranial roof.
60 Dermoid cyst.
61 Metastases of a cancer of the thyroid gland.
62 Healed trepanation opening.
63 Diagram of a Neolithic skull.
64 Radiograph of a trepanation opening.
65 Pseudo-trepanation. Peruvian skull with a big fistula as a result
   of pansinusitis.
66 Front view of a Pre-Columbian skull with five trepanation holes.
67 Peruvian mummy with a trepanation hole.
68 Trepanations made by a sawing technique.
69 Flint compasses.
70 Skull with enormous trepanation.
71 Skull with sincipital-T.
72 Ofnet skull no. 21.
73 Reindeer standing over a pregnant woman.
74 Dr Gaussen showing his Magdalenian hut.
75 Corset of bark.

# Foreword

Human life is short. Man is small. He discovers the vastness of space through the objects around him. He supposes that time is infinite. The lack of accordance between the length of his existence and of the past and the future, stretching to eternity, is the origin of his philosophy. Yet his thinking in space and time is two- or three-dimensional. If he looks for the end then he goes impulsively back to the beginning, about which he has certain assumptions because of written and oral traditions. He is aware of an evolution. This is not a constant process: there are ups and downs which confuse the mind and cause false conclusions to be drawn. The latter, supported by centuries of tradition, call forth critical judgment. It is this criticism which compelled mediaeval man (now purged), to go back to the source: here is not so much the rebirth as the birth of Free Inquiry.

This critical spirit has permeated all levels of science and stood out against dogmatism. It carried with it every scientist – clergy as well as lay. The centuries of stagnation at all levels, a result of the flowering of Christianity, were in principle against free inquiry. Medicine perhaps suffered most from this. Nevertheless it was the monks of Monte Cassino who, without hesitation, provided the necessary documents from their library for renewed study, which reached its culmination in the work of the Belgian Vesalius in the sixteenth century. It was a Swiss doctor from Zurich, J-J Scheuchzer, who claimed to have found *Homo diluvii testis* among certain fossils. We cannot accuse these deeply religious men of apostasy: they tried to find evidence of the Divine Word in a way typical of their age, and reached the inevitable conclusion: the division of belief and science, so that each might supplement the other.

Scheuchzer's (unknowing) mistake is typical and illustrates what might best be put in the form of a proverb: 'Today's truth is tomorrow's mistake, today's mistake is tomorrow's truth.'

We can only regard as laudable the Irish bishop James Usher's modest attempt at working out, in the 1650s, the date of the creation of Adam as 22 March, 4004 B.C. Darwin too had a modest nature, though it is difficult to believe this of the man who worked out the idea of evolution: that over the countless years more developed types had evolved from primitive forms. Where was creation now? We must remember that Lamarck, his French precursor, had his theory quashed by the brilliant pseudo-arguments of Cuvier. We know now that neither Lamarck nor Cuvier was right. Even Darwin's work has only a relative value, otherwise we should never have found it necessary to speak of mutation theory.

With the work of Darwin, and even more of his fervent supporter Huxley, dogmatism received a severe blow, in spite of the fact that Samuel Wilberforce, bishop of Oxford, thought it necessary to ridicule the evolution theory by asking whether Huxley claimed descent from a monkey on his grandfather's or his grandmother's side. With this the Gordian knot was cut, and the way was open for investigation of the past to become a separate discipline – the study of prehistory. To describe the lack of understanding towards the first investigators would itself be sufficient material for a book: we need only mention the tragedies of Boucher de Perthes and De Sautuola in the field of prehistory, of Dubois and Fühlrott in the field of anthropology. Fortunately such injustices have been made good in our day, and figures such as Emile Cartailhac, and Rudolf Virchow are still great: Cartailhac because of his *Mea culpa of a Sceptic* to De Sautuola, who discovered the Spanish glacial era art in the cave of Altamira, and Virchow – prince of anatomo-pathology (although he described the Neanderthal skull as 'diseased').

We are no longer so prejudiced, and have become so careful in our interpretations that we hold on to absolutely everything that we recover from a prehistoric site, so that something is still around tomorrow even if it seems without value today. Furthermore, investigators today are not only laymen but often also priests. We may mention Carvallo in Spain, Breuil in France, Obermeyer in Austria. Some of them were regarded as cranky when starting their difficult work years ago. Now their work has borne fruit: a solid construction has been erected on the basis of their ideas – the study of prehistory, which nowadays attracts all levels of society.

This discipline is one, however, which has to appeal to other sciences – prehistoric industry implies a prehistoric person who is studied by anthropologists. The anthropologist may also be a doctor, who will be tempted to interpret anatomo-pathological deviations among the pieces which he is given to study. Thus, together with prehistory and anthropology, palaeopathology came into being. It is hardly necessary to point out that the geologist, who studies the layers in the earth, the botanist who analyses pollen, the physicist who performs the C 14 test, have become essential to the full investigation of archaeological excavations.

For this reason I have presumed, in the course of this text, to give some consideration to and short explanations of these connected sciences, so that the reader need not consult special books in order to find his way in the labyrinth of time and cultural stages.

While reading the work of my colleague Dr Charles Morel on palaeopathology among his fellow countrymen I was struck by the respect the writer shows towards his predecessors, such as the great Prunières, whom he calls the first palaeopathologist. I feel a similar regard for Belgium: were not the finds at Spy decisive in determining the existence of the separate race of Neanderthal men? Did not Schmerling, professor of the university of Liège, devote a book to the palaeopathology of prehistoric animals in 1833, at a time when prehistoric man was still not regarded as 'real'? Schmerling was also repudiated by his contemporaries – themselves the victims of the dictatorial Cuvier. But today Schmerling's works are found to be of scientific value, even though conditions such as avitaminosis, for instance, were still discoveries for the future. It is no more than just to place Schmerling in the foreground and give him the attention he deserves: after all, he won his spurs in palaeopathology earlier than Prunières.

Not that I wish to judge Prunières in any way. On the contrary he is and will remain a great figure: great in life as a palaeopathologist, and great too as a doctor in death – he froze to death in the course of carrying out his work.

Not only Prunières ended his life in an unnatural way. Fate does not seem to have been very kind to palaeopathologists: Ruffer, the father of Ancient Egyptian pathology, died in a shipping disaster in 1917. Moodie, one of the greatest figures in palaeopathology, died from complications after a fall on the patio of his house.

I wish only that this work may contribute to a better knowledge of palaeopathology in particular and prehistory in general. If so I shall have reached my goal.

P. A. JANSSENS
Santander (Spain), August 1958
Antwerp (Belgium), October 1965

NOTE    Figures within parentheses in the text, for example (37), refer to the Bibliography, pages 159–165.

# CHAPTER 1

# Definition

Palaeopathology is the study of disease in the past, and in particular of those times for which there exist no literary sources. In 1914 Sir Marc Armand Ruffer defined it as 'The science of the diseases which can be demonstrated in human and animal remains of ancient times'. Ruffer was only concerned with Ancient Egypt, and for this reason Moodie extended the formula to include 'not only the diseases on the mummified animal and human remains of Egypt, but those of prehistoric man and fossil vertebrates as well'.

This study of disease will entail our tracing back the phylogenesis, from the existent *Homo sapiens recens* in the Quaternary, to as far back as the margin of the Silurian-Cambrian periods in the early Palaeozoic, something like 450 million years ago. At this point in time the first vertebrates appeared on earth (Table I) (37). Their inner skeleton allows us to detect pathological aberrations with certainty. But even the invertebrates, which existed before these, have not managed to keep their sufferings entirely secret from us, because of fossilization in either positive or negative form. A positive fossil is formed from those organisms where a hard casing has provided an outer skeleton; a negative fossil is formed in a way similar to that in which lava from Vesuvius fossilized humans and animals. In this way extinct soft-bodied compound-celled, and even single-celled, organisms have been preserved for modern science. Still further back we reach the Archaicum, from which period we are familiar only with bacteria, one of the most important agents in pathology today, but which we can hardly imagine without their almost necessary host, either animal or plant. Palaeopathology, therefore, takes us back to the origin of life itself.

Moodie (124) believes that parasitic diseases result from the antagonism between two forms of life. While there is no reason why this should not have happened in the Archaicum, present-day finds suggest only a harmless antagonism even at quite an advanced Palaeozoic stage. Hence he concludes that possibly the first fauna were free from disease. I agree that this might have been true of infectious diseases, but the supposition does not rule out functional diseases. Parasitism exists from the Devonian onward. Injuries do exist before this period, but not necessarily of a parasitic nature.

It was the intention of Vuillemin to extend the study of palaeo-pathology still further and apply it to plant-life. This extension he called 'Palaeophytopathology' (135).

From the Devonian period onward we find bacteriological and fungus activity and this activity increases considerably in the Carboniferous period. Before these periods material for study is scarce. Moreover we have to bear in mind that such activity is not necessarily a pathological process, but may also be a decaying process after death.

The study of palaeopathology should not always be descriptive. It should give us the answer to certain questions: is disease in-variable? Have the same diseases always existed? Or more ex-plicitly: have some diseases disappeared and new ones come into being?

Pathogenic factors work on the affected organism not only at a reactional or somatic level, but also in a psychic, sensitive way, when they will also be above the threshold value. Not all living organisms are aware of the pathological condition in which they themselves exist. But once this threshold value is surpassed, so that the conscious sensation recognizes or is sensible of a pathological condition, then the organism will try to promote the spontaneously arising reactions which will lead to stabilization and disappearance of the abnormal condition. In this lies the origin of medicine.

As norms have differed from one time and place to another, depending on existing philosophical, political, economic, and social currents, we shall assume a definition of the term medicine which will have to fit any period whatever, from the first human, whether Pithecanthropus or a Neanderthal, to the first barbarian who looked in wonder at the glittering armour of Rome's advancing legions. Indeed it will even have to apply to the sick of today, in the newly dawned atomic era.

## DEFINITION OF MEDICINE

What is medicine? – or rather, when can one speak of medicine? Nowadays medicine is a well-defined notion: it comprises every

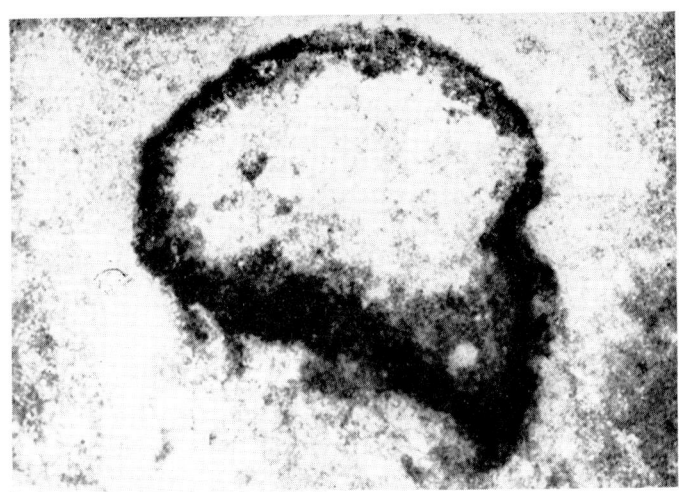

1. Cranial silhouette with face from a bell beaker grave at Schayk (Holland). From De Laet and Glasbergen, *De voorgeschiedenis der Lage Landen*, Pl. 18.

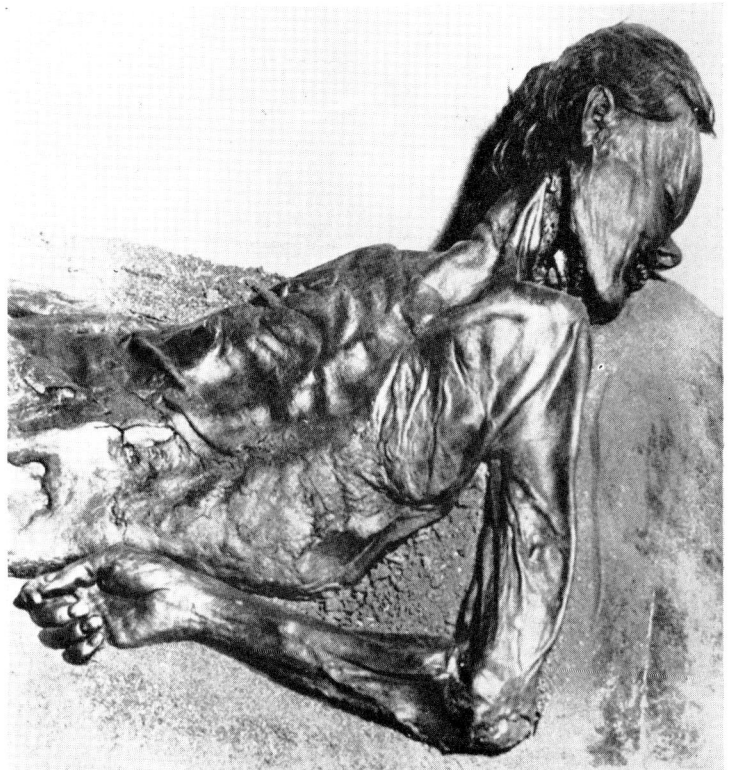

2. 'Bog corpse', man of Grauballe, Denmark, third to fifth century. The throat has been cut. From Wells, *Bones, Bodies and Disease*.

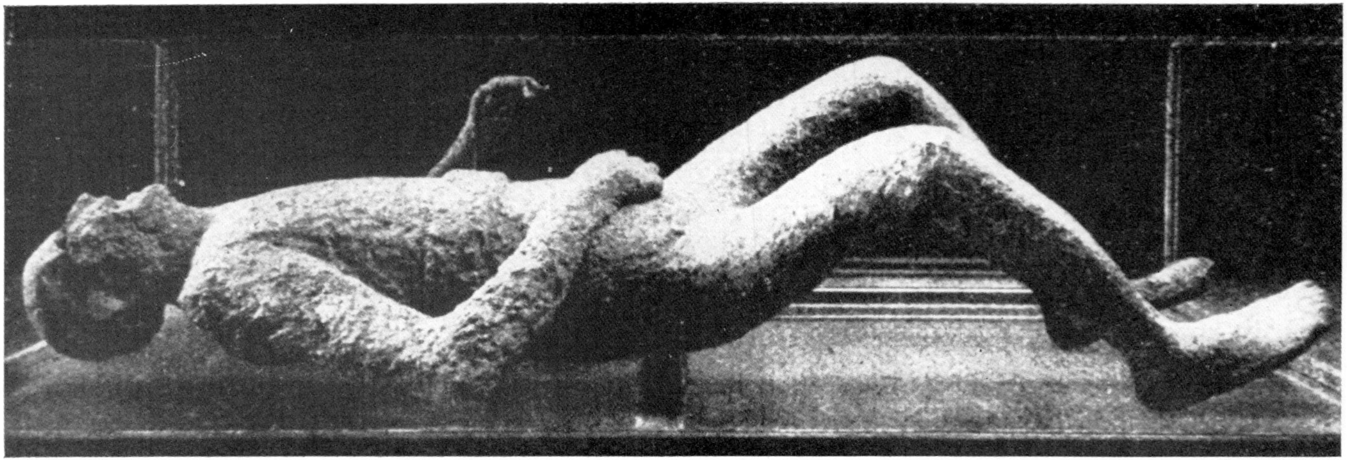

3. Plaster cast of a Roman, victim of the Vesuvius eruption. (A.D. 79.)

4. Flint knives, used by Egyptian embalmers. Photo: Ciba-archive.

action carried out to safeguard man against disease. These may be the strengthening of health through physiotherapy, or the protection of health against direct or indirect threatening influences through hygiene and social medicine, or finally action directly promoting recovery from disease by qualified persons with a minimum of knowledge required by law. These actions do not necessarily meet the standards set by an exact science: if empirical methods in medicine have decreased due to progress in the latter, they have not disappeared altogether. Until recently, for example, a doctor might prescribe sodium formiate as a substitute for sodium chloride – ordinary table salt – when a diet without salt was necessary, because he thought that the chlorine ion could harm a patient with high blood pressure. Now we know that it is precisely the sodium ion which is to be avoided. This, of course, does not mean that the doctor in question was a 'quack'!

It is apparent then, that simply the intention of helping the diseased is not sufficient to define the modern idea of medicine: a 'science-factor' is indispensable, and a 'law-factor' is necessary not only to protect the diseased, but also to protect knowledge. This has not always been so: a medicine-man who administers cinchona to a sick tribesman seems, to our way of thinking, to perform a real medical deed, because we know the working effect of quinine. If after this he exorcizes 'evil spirits' we do not regard this as a medical deed. In the eyes of his tribesmen, however, it is one. Thus we must accept that medicine in antiquity comprises all deeds which, in the eyes of contemporaries, aimed at promoting those methods that we now accept in modern medicine.

B

CHAPTER 2

# Influence of climate, economy, and environment

If Hippocrates considered the influence of climate and economy sufficient to devote a book to it, then this is certainly true of pre-historic times, when glacial periods alternated with warmer inter-glacial periods. Pathology of these periods clearly bears their stamp. Just as important is the transition from a hunting to an agricultural way of life. In our regions this revolution, the greatest that mankind ever knew, took place in the fifth millennium B.C. when agriculture and cattle-breeding penetrated into Europe via Asia Minor and Thessaly (148). This will be emphasized in due course.

Environment has easily the most important influence. C. Wells says: 'The intricate relationship between a people's way of life and the disease they endure is the chief reason for the study of palaeo-pathology' (191). This relationship does not have to be of a strictly pathological nature. It can also lie in the border area where anthro-pology and palaeopathology overlap. Consider for example the third facet on the ventral side of the ankle joint that develops through the crouching attitudes of primitive peoples, or the lateral curvature with pronounced tuberositas deltoida of the humerus of peoples who used the sling as a weapon. Furthermore it is not always possible to trace real pathological conditions. We can easily imagine that the Neolithic miners from Spiennes and Obourg, hacking flints out of the limestone day after day with their stag-horn pickaxes, suffered from silicosis. We might regard this as the first occupational disease, as we should also like to call the miner who met with an accident in the same flint quarry the first victim of a fatal work accident. Both conditions indicate a specialization of labour which really existed: first the mining and then the shaping of the roughly hewn axes, which only later got their final shape through burnishing. Occupational diseases and work accidents are among the products of civilization. In our modern society they even

4

require special legislation. We have only to think of the many possibilities of acute and chronic toxic absorption in our modern chemical industries, or, further, of accidents from electrocution, from explosions, or from working with compressed air.

## RELATION WITH RELIGION

The relation with religion is another dominant factor, as it still is today. Indeed 'philosophy of life is born of wonder about the phenomena of life in its endless variations', says L. Elaut (47). 'It comprises the philosophy of normal life, which the subjective consciousness calls health, and the philosophy of abnormal life, which the subjective consciousness or objective perception calls disease. Philosophical thinking about the origins and causes of diseases is medical thinking.'

This parallels the history of thought in general.

One tends to mention the religion and medicine of any primitive people in the same breath. Yet medicine precedes religion. The preservation of life, whether generally or individually, fulfils an inborn need or instinct. It finds expression within the framework of medicine before any notion of religious feeling, since sickness has to exist before it can be looked upon as the punishment of an angry deity. In the latter instance, however, a priest or the equivalent will be of more service than a doctor, unless the priest is at the same time a doctor. He will be at a higher intellectual level than his fellow tribesmen and this gives him many advantages in the study of the extending, though as yet ill-defined and undeveloped, fields of science in general. It is striking what importance the primitives attached to casual factors: can illness be a straightforward affection, or is it divine vengeance?

We have, as it were, the first application of the saying: 'Qui bene diagnosticat, bene sanat'. Moreover, interest in the cause seems almost greater than interest in the therapy. A further point of great importance to us is the primitives' highly developed sense of spiritual purity. On the temple of Aesculapius in Epidaurus was inscribed: 'Pure must be he who enters this temple; purity is to have nothing but holy thoughts.' This spiritual purity must however be demonstrated physically through taking baths and wearing spotless clothes (161).

It is easy to appreciate the importance of religious interpretation of disease, which will determine whether the attitude of society towards the sick is hostile or not. We might almost say that its importance was more prognostic than of real therapeutic value.

CHAPTER 3

# Definition of disease

It is not easy to give a definition of disease. According to Virchow disease is nothing other than to live in altered conditions. This definition includes injuries in as far as they are a cause of more general phenomena. In Bouchard's opinion disease is an abnormal functional condition of the organism. The phenomena which occur under this heading consist of the sum of disturbances which are directly or indirectly brought on by a pathological cause and of the reactions in the body which work against them. In palaeopathology one can formulate the condition 'disease' as a deviation from a healthy or normal condition of the organism which has left visible traces in its fossilized or mummified remains. From the preceding definitions we can assume that disease is as old as life itself; and further, if we agree that the preservation of life proceeds directly from an inborn need or instinct, we can deduce that medicine is just as old.

Even though the objects to be studied in palaeopathology are mainly 'documents' from the soil, we still need to make a distinction between primary and secondary objects. By primary objects we mean those anatomical-pathological remains which can be studied directly; usually they consist of bones from an inhumation. The body – regarded as document – may be found in very different circumstances by archaeologists. Indeed in the case of inhumation the bones will not always be found. This is primarily due to the physical and chemical effects of the soil. Calciferous soil preserves bones very well, while sandy soil may cause them to disappear completely. Nevertheless in the latter case it is still possible to take observations: with careful scraping of a dissolved inhumation the profile of the body will often be visible. In some cases it is even possible to observe the facial features (Pl. 1). Hard concretions such as gall-stones and renal calculus will be preserved. In other cases, as when the finds are recovered from bogs, the soft parts of the body

6

will remain; the acid peat soil having tanned, as it were, the skin and other soft tissue, although bones have disappeared almost completely through decalcification. In such instances preservation can be so good that when the 'Tollund Man' (third to fifth century) was found it was thought to have been a recent murder (Pl. 2). A further case is the preservation of the body in a negative form. Here we are referring to those victims of the eruption of Vesuvius who perished completely in the layers of ash at Pompeii, their bodies leaving a mere cavity (Pl. 3). Casts made through filling these cavities with plaster reveal a sinister picture of death from asphyxiation.

A further factor is the burial rite itself: cremation was a common form of burial which in Belgium predominated from the Middle Bronze Age down to the Roman period and which was still practised as late as the sixth and seventh centuries, at the height of the Merovingian civilization.

All religions agree on one point: the immortality of man and his soul. If this immortality is conceived as a physical rebirth, the body of the deceased will be bestowed special care, naïve as this may seem to us. Typical are the mummies, where more care has been given to the case of the body than to the vital organs. Mummification can occur either from desiccation in the burning heat of the sun, drying out before decay has time to set in, or from the well-known process of embalming. Other methods are also known, such as smoking the corpse and embalming in honey. Sometimes embalming involves the removal of the bones. An example of this is the tsan-tsa or shrunken head of the Jivaro Indians of Ecuador. Embalming sometimes precedes cremation: such is the custom with the Khasi of Assam. This brings us to what is called the burial of second degree: after the corpse had been subjected to a first treatment such as stripping the flesh from the bone – performed either by man himself or through exposing the corpse to putrefaction and the gnawing of beasts of prey – part of the bones was buried separately. It has been supposed that this might have been the practice during the Neolithic period in Belgium at Spiennes (175). Nevertheless I feel very sceptical about this assertion, because the skeletal remains were found in old filled-in mineshafts in which altered geological conditions may have caused part of the skeleton to be dissolved by water charged with carbonic acid seeping through, so that only certain parts have been preserved. This argument is borne out elsewhere in the action of our very porous sandy soils. Within one and the same Merovingian inhumation a skull might be found in a more or less well-preserved condition while the rest of the skeleton

7

has disappeared, or the legs might be preserved while the rest has disappeared.

Preservation of the body through freezing is also possible. The best-known examples are the mammoth finds in Siberia, whose flesh, after thousands of years, was still edible. A few cases of human remains are known, preserved in the ice in the high regions of the Andes and in the Polar regions, and especially in proto-Scythian burials in the Altai region of Siberia (191).

Secondary documents consist of pictures, sculpture, pottery, masks, and last but not least written works (190).

Paintings of mutilated hands in the cave of Gargas have long been regarded by archaeologists and ethnologists as mutilations resulting from some ritual practice. I have proposed elsewhere that this might more fruitfully be treated in the light of palaeopathology (75); I shall discuss further examples below. In this category come sculptures such as the steatopygous 'Venus' figurines or the bronze sculpture of a dwarf, an example of achondroplasia, from Nigeria. Masks can be treated in a similar way. As articles for everyday use, especially in the tropical regions, they often portray skin diseases, which occur frequently in these regions. More familiar is the pottery from the Mochica and Chimú periods of Peru, decorated with so many scenes taken from daily life. Here the potter has been affected so deeply by pathological conditions that his agile hand has seen fit to record the sharply observed picture in eternity.

The most important source however is written evidence.

# CHAPTER 4

# Chronology of the Quaternary

We have already defined palaeopathology as the study of patterns of disease in a period which lacks written evidence (writing first came into practice about 3000 B.C. in Mesopotamia). As this period has been best studied in Europe, especially Western Europe, most sources relevant to palaeopathology will be found here, where the period lasts until the arrival of Julius Caesar. Thus as well as the Palaeo-, Meso-, and Neolithic periods, it also includes the Metal era down to the Iron Age.

In this way we can follow the course of disease millions of years before man's arrival. We assume that Cro-Magnon (= *Homo sapiens*) man arrived 60,000 years ago and Neanderthal man 150,000 years ago, but dates vary a great deal according to the author: according to A. Senet (158) and H. Kühn (102) *Pithecanthropus javanensis* lived more than 300,000 years ago, while in the opinion of Moodie, over 500,000. Van Königswald (179) claims that the *Sinanthropus* (=P. pekinensis) first appeared more than 300,000 years ago, while Senet estimates 100,000. Heidelberg man lived more than 500,000 years ago according to Kühn (102), and Andérez (2), and in the opinion of Roy Chapman Andrews more than 750,000 years ago (3), (Table II).

At any rate we see that all the earliest human forms appear in the Quaternary. This period is divided into two major phases, the Pleistocene and the Holocene. The Pleistocene saw four glacial eras, in Europe called respectively the Günz, the Mindel, the Riss and the Würm eras. This phase ended about 8000 B.C., after it the Holocene begins. The Günz and Mindel eras are hardly known at all. We know a little more about the Riss, and only about the Würm are we reasonably well informed. This last is also the most important because in that period *Homo sapiens* appears. His cultural stages are also reasonably well known, and since a few years ago it has been possible to draw a more exact time scale with the help of

9

research into the radioactive Carbon 14 atom. The four glacial eras alternate with interglacial periods marked by a warmer climate, without ever becoming tropical, although animals typical of a tropical climate lived on our continent.

We can say that during the Pleistocene three different types of man succeeded one another, and, from a cultural standpoint, possessed three different types of industry. On the other hand they had certain 'negative' qualities in common: they did not know the use of metal, pottery, or weaving.

Flint, either with a cutting edge or pointed, was the main material for implements. As the implements develop, so we can speak of cultural phases. At first we notice a very slow evolution: one period may last tens of thousands of years. As we progress in time so this development moves faster, almost according to logarithmic values. In broad outline we can discern the following phases: first the Abbevillian, when man used flint blocks from which he chipped flakes to give a milled cutting edge. This type of hand-axe might have been characteristic of *Pithecanthropus* although so far we have no absolute proof of this. The next phase, in which stone objects were more cultivated, being flattened and more almond-shaped, with the outer crust nowhere visible, is called the Acheulean culture.

Then appeared a different species – Neanderthal Man. Through experience he was led to the discovery of a new technique in knapping stone. He reached the stage of making triangular points, blades, and more or less circular discs. Called the Levalloiso-Mousterian technique, this lasts until the last glacial era. Nevertheless certain groups seem to have realized that the use of flakes was very important, in other words that a flake could be more efficient than a clumsy core hand-axe. These cultures run more or less parallel with ones already mentioned and are termed the Clactonian and the Tayacian.

It is in the last glacial era that vast changes come about in every field. Neanderthal Man made room for *Homo sapiens*, who, as we know, is physically no different from man today. At the outset he gave less attention to flint and instead used new raw materials: horn and bone. From these he made arrow and spearheads, drills and needles. But to be able to make objects from horn and bone, man had to adapt his flint tools. This first phase of the later Old Stone Age, also called the Upper Palaeolithic period, was originally known as the Aurignacian. However more recent discoveries from excavations enable us to distinguish three distinct levels: the Old Perigordian – for which the Chatelperonian points are typical, the Aurignacian, and the Late Perigordian – for which the Gravette points are typical.

Carbon 14 dating has been applied here (using the known 'half life' of this atom), from which the following dates were ascertained for some well-known sites: Arcy-sur-Cure, 31,500 B.C., belonging to the Old Perigordian; in the Aurignacian we find at La Quina, 29,000, and again at Arcy-sur-Cure, 28,375 B.C. Abri Pataud, with a culture belonging to the Late Perigordian, was dated between 22,600 and 21,600 B.C. In general these periods are characterized by a regression of the stone culture, but in some places we notice a distinct revival and a climax in the working of stone, which contrasts sharply with the previous period and the one following. This is the so-called Solutrean, stemming from a mysterious people of whom we still do not know the origin. The earliest phase, the proto-Solutrean, was dated in Abri Pataud at 19,500 B.C. and in Laugerie-Haute at 19,735 B.C.

The following period is marked by a new regression in stonework and an advance in bonework. The manufacture of harpoons heralds the approaching end of the glacial era, because an increase in fishing is related to the rising water level in rivers. This period, the last of the glacial era cultures, is called the Magdalenian. In Lascaux the Magdalenian III has been dated at 15,000 B.C. During excavations in the cave of El Juyo, we were able to conclude that the Magdalenian III which exists there dates from 13,900 B.C. and the same culture in the well-known cave of Altamira from 13,500 B.C. The Magdalenian VI in La Vache cave was dated at 9700 B.C.

Here we reach the end of the last glacial era. Several changes obtrude themselves upon the cultures, as the glacier retreated northwards. These changes are evident especially in Belgium and the Netherlands.

With the transition from the glacial era to the Holocene there came into being a group of cultures which we call the Microlithic, when prehistoric man made very small flint implements, usually of geometrical shape. In Spain and France the first phase of these cultures is called the Azilian. In Spain, however, this period dated from the glacial era, while in France it was post-glacial. It is no longer possible to regard this transition as a separate culture: we assume rather, that it was a last phase of the Magdalenian (7826 B.C.). In the north, between 13,000 and 8000 B.C. we find a number of cultures which were in fact a continuation of the Magdalenian and the Perigordian: they are the Hamburgian, the Tjongerian, and the Ahrensburgian, as well as the Cheddar and Creswell cultures belonging mainly to England (174).

When, from about 8000 B.C. onwards, during the preboreal climate, the temperature became moderately warm, cultures of the Ahrensburg type predominated. From about 6800 B.C., during the warm and dry boreal climate, a new culture, the Tardenoisian, spread over Belgium. The earliest phase is dated at 6012 B.C. and the latest at 5252 B.C. In the Atlantic phase which followed, there appeared in the north cultures of Mesolithic character like the Tardenoisian, among which some, such as the Maglemosian, foreshadowed the Neolithic period. This next period, the Neolithic, is sometimes referred to as the period of polished stone. Its breadth and importance lie not so much in the method of working stone, as in the far-reaching economic revolution it brought about. The hunter became farmer. It is as if the human mind after slumbering for countless centuries has only now fully unfolded. One invention followed another, and it is strange to realize for how short a time the different phases of the Neolithic lasted; they seem to be only a short flaring up compared with the Palaeolithic cultures. Furthermore they were soon overrun by the Bronze Age cultures, which themselves gave way relatively quickly to the Iron Age. This brings us to historical times.

Civilization is geographically discontinuous. We are already living in the Atomic Age, while some tribes in South America and Australia still live in the Old Stone Age. When our regions were in the Middle Bronze Age, Egypt already had a solid civilization. Still, commercial relations existed between the two, as is shown by the string of beads of Exloo-Odoorn, consisting of earthenware beads of the XVIIIth and XIXth Dynasty (c. 1400 B.C.) (116). The rich copper deposits in Spain, the basis of the Almerian culture, attracted the Aegean peoples, so also did amber from Denmark (32). Although Belgium, being poor in raw materials, contributed little or nothing to the economic network it did not remain totally cut off from outside influences. Strictly speaking, Egyptian medicine does not belong in a study of palaeopathology. However, I shall still refer to it in connection with anatomo-pathological remains from the corresponding period (146). It is a supplementary source because the manner of burials in Egypt was such that it has left behind anatomo-pathological material. This is not so for Spain, where in Palaeolithic times cremation must have been prevalent (24). The Pre-Columbian period in the New World is also within the scope of this subject, even though dating is particularly difficult. Because burial customs remained the same after the arrival of Columbus it is not always possible to determine whether the remains belong to the Pre- or Post-Columbian period. Examination by means of radioactive Carbon 14 will be of crucial importance for this period.

Sometimes mention will be made of pathological remains from the first periods of the formation of the earth's crust. The Secondary era, characterized by the enormous development of reptiles, is a rewarding period for indicating certain patterns of disease. I refer to these only to give some idea of the vast time scale within which we can discuss the earlier occurrence or non-occurrence of a certain pathological process.

Since the study of the material left by Egyptian embalming is so important, I should like to draw attention to the fact that, although the Egyptians knew iron, embalming was performed with the help of flint knives (Pl. 4). This is recorded by Herodotus (30; 123), and is the ritual continuation of the use of a previously common implement. The same can be said of the Jewish people, for the Bible says: 'At that time the Lord said unto Joshua: Make thee sharp knives [of flints] and circumcise again the children of Israel the second time' (Joshua v. 2).

It is obvious that we shall find parallels between primitives of today and prehistoric tribes. Nor have we, living in modern times, managed to rid ourselves completely of primitive customs, even if these only linger on in the form of superstition. Children in Calabria, for instance, wear necklaces of pierced animal teeth to prevent complications during teething. Excavations from the upper Palaeolithic period reveal quantities of such teeth. We might therefore doubt that these teeth were worn for purely aesthetic reasons, and assume rather that they were worn for medico-magical reasons (22). In Toledo, a province of Spain, people nowadays still wear small goat's horns round their necks. The horn is set in a metal pendant, provided with a loop. It protects the bearer against disease due to the 'evil eye'. In passing I should like to suggest a comparison between these articles and the pierced *bâton de commandement* from the Upper Palaeolithic Period, for which so far no satisfactory explanation has been found (Pl. 6).

# CHAPTER 5

# Material

It has become a habit to compare the life and customs of Palaeolithic man to those of primitives today, and even to deduce the one from the other. This is a little like building a house on sand. Contemporary climate and tribal differences have at any one time such widespread interrelated effects that such comparisons are hardly applicable in palaeopathology. Further, Cartaillac says: 'One has to take into account human freedom of choice'; and De Laet warns: 'Une similitude technologique n'implique pas nécessairement des institutions sociales ou religieuses identiques (32).'

Since the material at our disposal consists mainly of bones, our first reaction is to imagine that we shall learn exclusively about certain specific diseases of the bone. This is not true at all. Furthermore, primary skeletal affections are rare. Carcinoma of the bones is usually a metastasis; osteomyelitis and—if we are to believe American research – rheumatic arthritis must be traced back to a focal infection such as tonsillitis. The problem is even more complicated when specific infections, such as syphilis, are involved. It is difficult enough to diagnose this disease with certainty in recent bones, let alone in fossilized ones, added to which it is no longer possible to trace the pathogenic factor as is the case with certain coccal infections. Further, a victim of syphilis need not necessarily show bone lesions. We talk after all of the neurotropic and dermotropic virus (the same applies to tuberculosis).

The problem of the history of syphilis has still not been solved completely; whether syphilis occurred in antiquity or was brought into Europe by the crews of Columbus's ships is another question to be answered by palaeopathology. A big handicap is that lesions caused by syphilis present a similar picture to those caused by the tropical disease framboesia, or yaws. It is also very difficult to decide whether an osseous deviation is the result of a pathological process or was caused after death through chemical erosion by soils, through rodents, insects, or saprophytic organisms. In this case microscopical investigation can be used to advantage.

A bone can show a disease indirectly. Radiographical examination of seemingly healthy bones may reveal diseases suffered early in life. Serious affections during youth, a period of rapid growth, cause scars in the bone, known as Harris's lines (185) (Pl. 7). These can also result from starvation. Systematic examination of these lines gives us an idea of the morbidity of a people without giving any specific diagnostic information. If more are found in women than in men we can guess at the patriarchal inclination of a primitive society. They can give us indications of the proportion of disease to health among populations who were hunters, farmers, or shepherds. We can also study the correlations between the number of Harris's lines and pathological conditions such as dental caries, or biometrical factors such as length-growth. This is still a fallow field which in the future may bear rich fruits.

CHAPTER 6

# The importance
# of palaeopathology

A final but no less interesting question is that of the importance of palaeopathology and the study of the history of medicine in general. The uniformity of the sources makes any introduction to palaeopathology itself very elementary. Klebs (61), who discovered the diphtheria bacillus, has said that the history of medicine teaches us the influence of epidemics on social revolutions, and of diseases on the inhabitation and desertion of particular regions, of diseases which attack in youth, and others which lead to sterility, leading in turn to the gradual extinction of a people.

Diseases certainly have caused the extinction of races (124) but it is impossible to find out about them because epidemic diseases usually do not cause bone lesions. The latter are more often the result of such chronic diseases as tuberculosis, syphilis, and cancer. It is not likely that virulence was greater in the past than it is now, otherwise survival would have been impossible. Nevertheless one has to be careful: it is not impossible that with the extinction of a race the cause of that extinction, in this case disease, has disappeared too.

It is also possible that a vanished disease may return after centuries, even if in a different form. Sweating sickness, or 'Sudor Anglicus', a mediaeval disease described by Hecker in 1846, might have reappeared in Europe in the first half of this century. The last well-known epidemic of Asian influenza, an epidemic which took the form of a pandemic, gives us an idea of the variability of virus affections.

While disease may have contributed to the extinction of the giant dinosaurs, other factors are also involved. Decline was gradual, so that the age of the race as a whole is a relevant factor. Moreover these creatures were extremely cumbersome: the largest weighed up to 39 tons and measured more than 60 feet, so they were not very mobile. Apart from this there was a disproportion between their size and their nervous system: their head was hardly the size of a horse's head and their brain was no larger than a fist. Climatic changes must also be considered, for the skeletal lesions are not the sort that may be counted among the causes of extinction. The rise of mammals must surely have been of primary importance in the extinction of large reptiles. These mammals quite possibly fed on the eggs and young of such reptiles.

The Enzypteridae, a group of giant spiders, appeared in the Cambrian period, assumed vast proportions during the Silurian and remained only in a few scattered forms at the beginning of the Carboniferous. They died out before they were known to have suffered any lesions. Weakening of immunity, as a result both of the ageing of the race and of the first appearance of disease, has had an important bearing on the extinction of that race. Trilobites are an eloquent example of this.

In 1855 Schmerling (155) emphasized the importance of palaeo-pathology, though he based it on an incorrect premiss. He thought that rickets was not a disease of civilization, because it had occurred in the cave bear. It was necessary to point out such things to assist contemporary anatomo-pathologists in their work and to get rid of a lot of wrong-thinking. We know now that the lesions were not due to rickets and that rickets did not occur in the Old Stone Age.

Palaeopathology is handicapped by the fact that some lesions alter the typical structure of a fossil, which is then discarded as worthless although it might be of great value to the anatomo-pathologist.

A man is usually buried near where he has lived, and his burial place has tended to attract the archaeologist more than the place where he lived. This is hardly surprising. In his grave his fellow tribesmen provided him with articles for the future life. And because this was supposed to be a new and better world, he was given the best and most beautiful objects. In contrast, those with which he lived have usually been found worn or broken. Even so archaeology, like all branches of science, is becoming more and more specialized, so that it can no longer embrace all related sciences. Archaeology, like medicine, involves team-work: geological investigation for geological dating, phyto-biological examination for pollen analysis, nuclear-physical investigation of Carbon 14 for a relatively exact dating. Finally there is the skeleton. Alas, in many cases so little attention is given to these pitiful remains of bones! And still the archaeologist appeals to the anthropologist and hopes to get a maximum of information about often badly treated pieces of bone.

Anthropologists cannot emphasize strongly enough how badly these remains are treated: for days a skeleton is left half uncovered, one part still in the damp soil, the other in the burning heat of the sun, while pottery which would withstand such treatment much better is taken away quickly for fear (a justified fear) of theft.

What about the palaeopathologist? It cannot be denied that his examination has often thrown new light on rigid and cliché-ridden descriptions of certain peoples. I established for example, that the Neolithic people from Furfooz were not peaceful, simple peasants who could only root around in the soil with digging-sticks, but, on the contrary, were notorious fighters. This could be deduced from their healed 'parry fractures' of the ulna of their forearms. The archaeologist who claimed that these people were cannibals because their bones showed certain defects, would have done better to submit the bones to a precise examination by a medical doctor, when his 'thriller' theory would have been discarded at once. What archaeologist can prove that the crouching posture, with the highly drawn up knees, in which skeletons of Palaeolithic and Neolithic graves are found, such as those from the caves of Grimaldi or from some 'Bandkeramik' burial places, results from the tying of the lower legs to the thighs and the body, rather than from extreme emaciation after a cachectic disease? (99) Archaeologically unindentified objects have sometimes been used for medical and para-medical purposes. For example I have proposed that a stiletto, found in a Gallo-Roman burial of a young woman, might have been used to procure abortion. The woman must have died during this operation, and the object was not removed, for it was found between the pelvic bones, at the height of the pubic bone, the point of the object pointing cranially. If the archaeologist then asks, with good reason, why this stiletto was not removed after the woman died, it is up to the palaeopathologist to show from the possible diagnosis of the cause of death as a simultaneous and sudden complication such as an air embolism, that her family and friends could not have been aware of what was perhaps a secret attempt at abortion.

For certain bone defects the archaeologist will not always be able to provide an explanation or diagnosis. For example, after a very carefully studied differential diagnosis of circular holes in a shoulder-blade, M. A. de Lumley (34) could only conclude that they resulted from an unknown disease, possibly an extinct parasitosis.

6. *Bâton de commandement* with engraved figures, made of deer horn. Prehistoric museum, Santander (Spain).

5. Goat horn, mounted in silver. Province of Toledo. Author's collection.

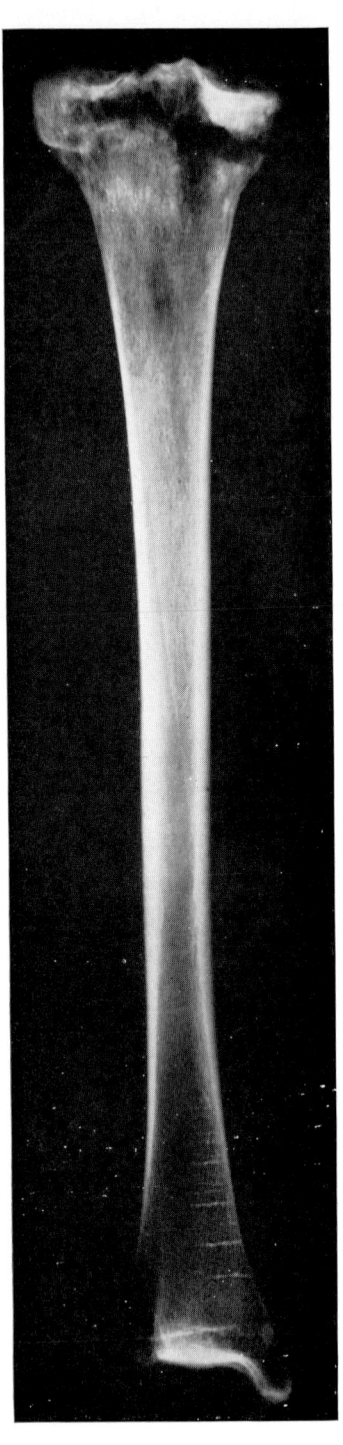

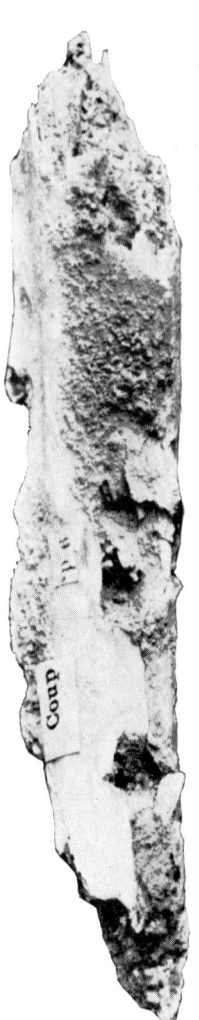

8. (above) Piece of bone from the burial cave of Furfooz. The two conical impressions are due to canines of a fox. The label, saying *coup* is visible.

7. (left) Radiograph of a tibia with Harris's lines in the distal part of the shaft. From Wells, *Bones, Bodies and Disease*.

9. Remains of cremation from a burial mound of Hogeloon, The Netherlands. In the centre a partly calcined flint point.

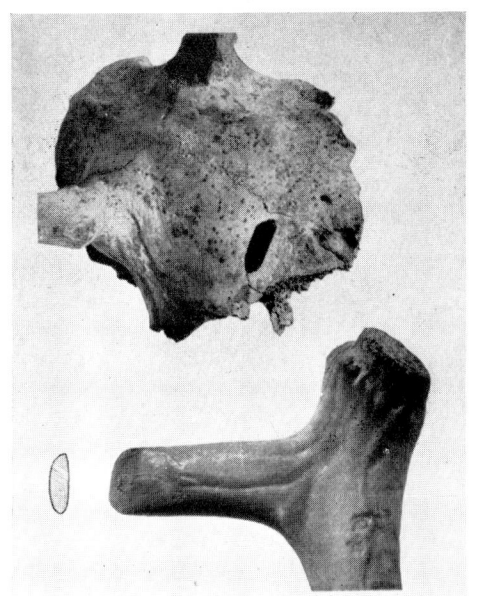

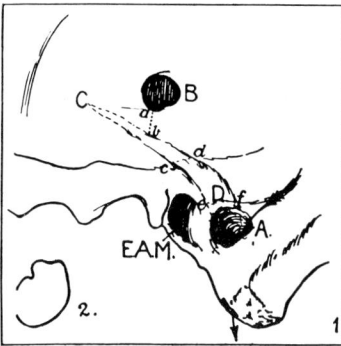

10. (left) Part of a reindeer skull with characteristic opening, caused by a Lyngby axe. From Rust, *Die Alt- und Mittelsteinzeitlichen Funde von Stellmoor*.

11. (right) Diagram of the lesions of a skull from Rhodesia. From Ciba Symposium, vol. 2, 1940. A and B: perforations. C–D is taken to be a gully, caused by drainage under the skin.

C

12. Skull of a man with three impacted fractures. From MacCurdy, *Human skeletal remains from the Highlands of Peru.*

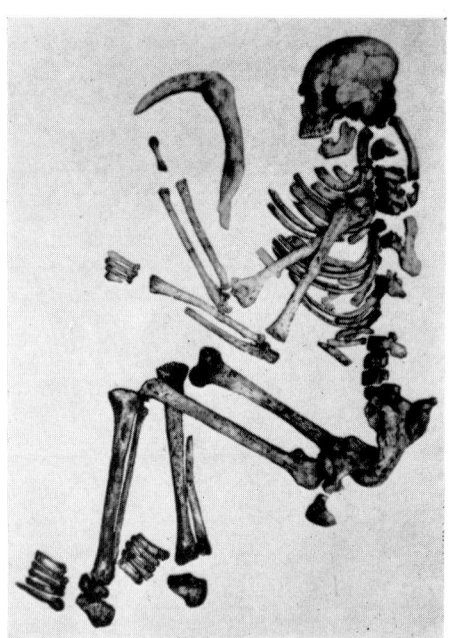

13. Skeleton of miner, who met his death still holding his deer-horn pickaxe. The man is of the brachycephalic type. Photo: M. Glibert, Koninklijk Instituut voor Natuurwetenschappen, Brussels.

# CHAPTER 7

# Skeletal remains

As skeletal remains are easily the most important source of evidence in palaeopathology I shall first draw attention to certain points which can lead to a false reading of the material under consideration. A bone can show lesions that have nothing to do with pathological or traumatic factors. After a completely normal burial, bones can be affected by plant roots, chemical or physical factors, and by insects and bacteria. Dripping water can cause very confusing lesions. The spade of the digger can inflict extensive damage, but in this case the patina of the bone will reveal the cause immediately. If descriptive methods do not satisfy, one can still have recourse to technical investigation, particularly to radiographical examination. Apart from this there is microscopic examination, but the results are not always conclusive. Pales (135), Sigerist (161), and Moodie (124) have mentioned a number of other techniques which deserve attention. Paring is the best method when dealing with highly fossilized bones for these are sturdy and very thin shavings can be made. Quaternary bones, on the other hand, are not suited to this method and even present great difficulties when other methods are used, for they often pulverize. Sigerist advises fixing the bone in formaldehyde (161). After it has been decalcified the bone may, according to the usual technique, be cut, coloured, and embedded in celloidin. I have seen quite good results from sections, prepared by my colleague Tverdy, where decalcification was carried out in a solution of 5 per cent trichloracetic acid. Pales (135) advises using phloroglucinol to decalcify bones.

When the deceased is interred a great many chemical factors determine the state of preservation in which the archaeologist will find these human remains centuries later. The burials in Denmark provide a well-known example, where bodies have been preserved very well in the peat, quite remarkably so in the case of the so-called

C

Tollund Man, already mentioned. When the putrefaction bacteria have worked on the soft parts of the body, chemical reactions will affect the bones if the conditions allow. Franchet (50) has emphasized the action of percolating water charged with carbonic acid in decomposing bone tissue, osseine as well as tricalcium phosphate. The latter is almost completely insoluble in distilled water (0·70 milligramme per litre), while water saturated with carbonic acid dissolves 87 milligrammes per litre. But in order to exercise its dissolving capacity fully, ground water must be in motion; this explains why bones can be found in perfect condition in boggy regions.

The bone may be partially or completely dissolved. But even if it is completely dissolved one can still find traces in the soil – the so-called corpse silhouette, where the features may be more or less recognizable.

Franchet has also studied the action of fire on bones, which is of interest when the remains of cremation are to be studied. He distinguishes eleven phases. In the first phase the bone takes on a yellow colour, in the second phase a light brown one, and in the third phase dark brown. In the fourth phase the bone is completely burnt and so is black, turning to indigo-blue in the fifth. In the sixth it is blue-grey. In the seventh phase, at 600°C. the bone turns white. In the eighth phase it contracts and becomes deformed. In the ninth splits appear together with torsion of the bone tissue and in the tenth vitrification begins, which gives the bones an appearance of porcelain. In the eleventh, with the temperature at 1,200°C. the bone melts.

Rodents and carnivores usually leave clear teethmarks on the bone. We could see this on the remains of bones at Furfooz (Pl. 8). Here these lesions were originally described as '*coups*', '*entailles*' and '*stries*' and were mistakenly regarded as results of cannibalistic practices (45, 143). It is much more likely that the two conical imprints on the piece of bone shown in the plate were made by the canine teeth of a fox (93).

However, we do have to take into account those lesions inflicted *post mortem* by man himself and connected with ritual customs. Following Rouillon (149) they may be divided into two groups: stripping the flesh from the bone, and breaking of the long bones.

Cleaning of the bones can be achieved in two ways: the first through ordinary decay of the soft tissue of the corpse, above the ground, aided by carnivorous animals which leave their teethmarks on the bone; the second by a man with the help of flint knives, which will also leave visible traces on the bone. After the bones have been stripped of their flesh they are buried or cremated. Burial of the second type explains why in some cases an incomplete skeleton has been found and why in other cases only one of two paired bones were present in the grave.

Rouillon has mentioned special preparation of the long bones in connection with their ritual breaking. First, notches are made in the bone. These occur at the line of fracture, and are always to be found on the diaphysis, never on the epiphysis, which seems strong proof against cannibalism, assumed by Le Baron: if this were the case it would have been the epiphysis which would be involved, 'parties les plus tendres des os, pour ne pas écrire les meilleures et les plus facile à consommer!'

According to Baudouin (11) only one long bone [for each person] was buried at Vaudancourt in France; at the same site he found only nineteen kneecaps for at least eighty-three individuals, a fact which he interprets as follows: during the process of desiccation the kneecap must have stayed attached to either the femur or the tibia by the knee tendon and afterwards have fallen off. A metatarsal bone of a child found in the vertebral foramen of an atlas seems rather poor evidence for the argument that an ossuary was only a burial place for bones which had previously been stripped of their flesh. For this fails to take into account the disturbance caused by burrowing animals. Not only were long bones broken, but skulls were also separated from the facial bones, in which case the skull shows signs of scraping. The cleavage is effected by first making a series of holes along the intended line of fracture. The same author also mentions the ritual fracture of a mandible.

Exactly how much importance we ought to attach to these facts is difficult to know. Are we really confronted by particular bones, or have the various burial places been put into use again after a certain lapse of time, as in Belgium? In other words, have the original skeletons only been partially cleared away? In Belgium this was certainly the case, as is shown by the burial cave of Furfooz (93); processes such as stripping of the flesh or ritual bone fractures do not seem to have been current. Besides this we know of only one case of post-mortem trepanation, in the burial cave of Hastière (Trou garçon). Rahir (142) has emphasized that he discovered no facts which could point to the ritual stripping of the flesh in Belgium. He usually found the skeletons in anatomical connection, as in the Trou du Blaireau at Vaucelles, for example. In cavity B of the Trou du Crâne at Furfooz also he found foot bones in anatomical context, and in the Trou de la Mâchoire he found a femoral head still in the hip joint. Also in the *marchets* he could always find an anatomical structure, although he sometimes speaks of 'bone packets'. Similar doubts also arise over remains of cremation burials. According to Rouillon nothing is left of the bones but ash after cremation of the corpse, though this would not be so if the bones were cremated after they had been stripped of the flesh, or if cremation took place in a very hot fire, when some of the bones would have kept their original shape. After examination of the cremated remains of the burial site at Hogeloon (14) I cannot agree wholeheartedly with this theory (Pl. 9). Among these remains was a burnt flint arrowhead, the likely cause of death, which must have remained lodged in the body during the cremation. Had the bones been bared of their flesh this arrowhead would not have been cremated with the body, even if it had penetrated a bone. Yet in these cremations we do find bone fragments, which would seem to contradict Rouillon's theory. True we find bigger cremated pieces, as in the urns found at Luiksgestel, amongst which parts of burnt diaphyses more than 7 cm. long could be recognized. Still there are no traces of the use of flint knives during the ceremony. So far neither inhumations nor cremation burials have provided any proof whatsoever of ritual baring of the bones among Neolithic and protohistoric peoples in Belgium.

The examination of cremated remains is a difficult task not only for the anthropologist but also for the palaeopathologist. But it is important for two reasons: first, the great number of burial places with cremations (79, 80, 82, 83, 84, 85, 86, 87, 88, 89, 90, 94); secondly, the fact that, with hardly a break, cremation lasted for more than ten centuries in these regions (31). In general the first cremations are found in the Neolithic period in the 'Bandkeramik'.

According to Carvallo there must have been cremations in Spain during the Palaeolithic because so few human skeletons are found in the culture layers, which are metres deep. In Belgium cremation was not replaced by inhumation until during the first period of Roman expansion, that is to say in the first century A.D. (89). Cremation was still in regular use, however, as late as the seventh century when the Merovingian culture had already started to decline (87). It is not at all surprising that many authors such as Fricke (51), Vlêck (178), Grimm (65), Wells (183) and Gejvall (56) have devoted special attention to this topic.

Apart from the cremated remains at Hogeloon already mentioned, where we can reasonably suppose that the cause of death was of a traumatic nature, cases of spondylarthrosis could be recognized in cremated remains from the urnfield of Neerpelt (80) (see chapter on Arthritis).

In the report on cremated remains from the urnfield at Grote Brogel (Limburg), dated to the end of the Iron Age, I wrote (90):

This is probably a man of more than forty. Attention should be paid to the wide meshes in the bone. Since X-ray examination revealed no sign of decalcification we might regard this as a case of osteoporosis. A closer examination shows that several of the flat skull bones are affected by this process and appear inflated, while others are normal. These signs are also apparent on fragments, probably from the pelvis, and one shows a woolly aspect. Nothing has remained of the vertebrae so no particulars can be given. Furthermore one cannot speak of hyperostotic zones because the fragments are too small. In any case it is not a senile process; it may be a presenile osteoporosis but the cranial signs of sex and age point rather to Paget's Disease.

This disease has been described by Pales in connection with skeletal remains from the Neolithic (see below).

In a comparative study of present-day cremations Wells (183) has stated that bodies of persons who have suffered from a cachectic disease, such as cancer or tuberculosis, are consumed by fire with more difficulty than are those of strong fat people suddenly killed by a coronary disease or brain haemorrhage: the presence of sufficient fat tissue might here promote combustion. On the other hand, bones of fat people tend to deform more than those of thin people during cremation. Wells has noted that the optimal cremation temperature lies between 820° C. and 900° C. At the same time he points out that during cremation the bone shrinks very much less than we should expect.

He also points out the presence of clinker. This is a brown mass, rarely larger than a hazelnut, interspersed with tiny cavities whose size varies between that of a pin head and a pea. The surface is shiny and glassy. It is found under the head and may represent the transformed keratin of the hair. However, this clinker is only produced in the presence of fat: burnt hair alone does not produce such clinker.

Furthermore, from his findings Wells deduces a connection between the body and the pyre. If the body was placed on the ground and the pyre above it the incineration of either the shoulderblade or its crista, or the sacrum and the nuchal part of the occipital bone was not complete. If the body was lying prone, incineration of the frontal bone, the facial bones and the medial parts of the clavicle and both patellae was less thorough. If the body was placed on top of the pyre we can expect a complete and intense cremation because the remains of the body sink gradually into the heat of the fire as the pyre collapses.

The difference in the quantity of cremated bone from one urn to the other is striking. In some cases the total weighs more than 1 kilogramme, while in others hardly ten grammes. In the first case we find practically every bone represented by some anatomically recognizable fragment, mixed with tiny fragments of bone and charcoal; it seems here that the burnt pyre has been swept together and put in the urn. In the second case the small amount of bone consists mainly of bits of diaphyses and the petrous bone, seldom flat skull fragments; it look as if these bone fragments have been selected after the pyre had burnt out and cooled down, and put in an urn: indeed these particular bone fragments are the easiest to pick up!

# CHAPTER 8

# Traumata

It is obvious that traumatic lesions form an important part of the study of palaeopathology. First, because bones are the most important material that has been preserved; and secondly because the struggle for survival demanded great and dangerous feats from prehistoric man (38). A fracture is an important stimulus which provokes a healing reaction in the endosteum and the periosteum with the formation of new bone. This new bone is usually produced in excessive quantities so that a callus is formed. In some cases the callus does not develop because of infection or too much movement, and a pseudo-arthrosis, or false joint, is formed. Such a reaction may be caused through the presence of a foreign body in the bone, such as a spearhead or an arrowhead. In other cases the influence of the trauma is not so great and a partial fracture or even no fracture at all is caused, but even then the working of the periosteum is sufficient to cause clear macroscopic changes.

The first description of abnormal bones was that by the German, Esper, in 1774, of the femur of a cave bear which showed a considerable thickening. This was taken for an osteosarcoma; in fact this lesion was callus round a fracture.

The oldest known fracture is of the radius of a Permian reptile, Dimetrodon (124). Callus round fractured bone is small in reptiles, for these heterothermal creatures can stay still for weeks without taking food; this is not the case with isothermal mammals, and the necessary movement when they are looking for food constantly irritates a fracture, causing the production of more callus.

It is quite logical that fractures of the extremities predominate in animals, except the Phytosaur, and more in males than females. Abel attributes this to fighting during the mating season, and in particular to bites and to blows delivered by the enormous tail. When man comes on the scene more fractures of the skull occur. Typical of these are the holes in the skulls of reindeer found by Rust (151). These injuries were caused by some sort of axe made from reindeer antlers with one of the branches sharpened (Lyngby type); with this axe injured reindeer were probably finished off.

Already, during the process of becoming human, man acquired the taste for killing his adversary with blows on the head. Some skulls of South-African Australopithecines show a distinctive form of fracture consisting of two depressions lying close together. These are considered to be the blows made with an antelope humerus, since its wide condyles correspond with the depressions (191).

The genus *Homo erectus*, which includes the well-known Java and Pekin groups, counts many members who died a violent death due to cranial injuries. In the first group, of which thirteen examples are known with certainty ('Catalogue des Hommes Fossiles', Congrès géologique international, Alger, 1952), several seem to have been murdered, although they may have been the victims of a volcanic eruption. Of the eleven skulls found on the Solo river, assumed to be intermediate in form between the Neanderthal and the *Homo erectus*, four showed head-wounds which had caused death. Of the second, the Pekin group, also referred to as *Homo erectus pekinensis* and found in the cave of Choukoutien, cranial remains of about forty individuals have been recovered. All showed head injuries with the head severed from the body. At least one of them had previously escaped death, for the skull showed a healed wound on the frontal bone. This hill does not seem to have been satisfied with the blood of these precursors of man, for in a later Palaeolithic phase a whole family consisting of four adults and three children was slaughtered with clubs and pointed weapons in the so-called 'Upper cave'.

Nearer to us is the cave of Ofnet (Bavaria) where the well-known 'skull-nests' of a murdered tribe have been found. Because of the special importance of the injuries on these skulls I shall refer to them at length below.

Such injuries of the skull are frequent in man. The skull of the Neanderthal shows a scar above the right eye-socket, and the skeleton also shows a serious lesion on the left humerus, as a result of which it had stayed weak. The proximal part of the left ulna shows a healed fracture of the olecranon, which had widened the fossa articularis (70).

A skull from Broken Hill, Rhodesia, shows a secondary infected lesion above the left ear. Mollison is of the opinion that this injury was caused by a bite from a leopard or hyena (100) (Pl. 11)

A skull from Obercassel belonging to one of the Cro-Magnon race shows a traumatic lesion of the left parietal bone (105) cited by Le Baron.

The 'injury' of the well-known skull of the 'Old man of Cro-Magnon' is not of traumatic or pathological origin, it is simply erosion caused by dripping water penetrating the burial cave. This post-mortem lesion consists of a circular depression 4 to 6 mm. deep and situated above the right sinus frontalis. There is no reaction on the face of the tabula interna. This skeleton also shows a perforating lesion of the pelvis, presumed to be the result of a stab wound.

The skull of the woman of Cro-Magnon, however, certainly shows a traumatic lesion (168). This consists of an elliptical opening 29 mm. long and 8 mm. wide. The walls are straight and steep and there is no reaction at the level of the tabula interna. There are traces of osteitis and hypervascularization present which, according to Nelaton, point to the fact that the woman may have survived her injury for twelve to sixteen days. The cause, according to Broca (105, 106), would seem to have been a short fierce blow from a stone axe.

During the Neolithic period cranial fractures frequently occur: a fracture similar to the one just described has been noted on the skull of a woman found at Sordes (Landes). The lesion is situated at the level of the right parietal bone.

Müller (129) has mentioned a Neolithic skull of a man aged between forty-five and fifty years old with four lesions, possibly all healed. The first is on the right parietal bone, an impacted fracture 28 mm. wide and 35 mm. long, probably caused by a blow from a blunt object. The cranium has been smashed inwards about 5 to 8 mm. and has healed. A small fistula and an uneven callus on the tabula interna could have been due to discharging. The second lesion is on the forehead, above the medial part of the upper edge of the eyesocket. It also is a depression 24 mm. wide and 30 mm. long; the bone has been staved in about 3 to 5 mm. There is little formation of callus, and it is not certain whether there was a fistula. The third lesion consists of the destruction of part of the lateral half of the upper edge of the left eyesocket 32 mm. long and 25 mm. broad. This lesion would have caused the loss of the left eye and may also have been the result of a blow from a blunt object. On the inside of the lesion callus has developed around the edge. The fourth lesion is also an osseous one on the right side of the upper jaw. This defect measures 20 mm. by 20 mm. The lateral incisor and the canine tooth have been knocked out, as well as the outer wall of the alveolus of the first premolar, which has fallen out post-mortem. The first incisor had been pushed to the left by the blow. Although the injury had probably festered it nevertheless healed. Müller mentions the possibility of a fifth lesion: a slight depression of the forehead on the left side just above the second lesion. The successful healing of all these lesions once more gives us an idea of the resistance of prehistoric man.

MacCurdy (113) has mentioned the case of a young man of about nineteen or twenty with three depressed fractures of the skull. He was found in Peru and was from the Pre-Columbian period. All three fractures seem to have been caused by the same club. The first one is situated between the lambda and the obelion, left of the sagittal suture. The other two, probably delivered by one and the same blow, touch each other and are situated on the occipital bone, underneath and left of the lambda. The injured man would have died before any surgical help could have been given. From only one fracture has the piece of bone fallen out, but this happened after death. From this opening a fissure runs over the angulus mastoideus ossis parietalis from the left parietal bone to the temporal just above the mastoid (Pl. 12).

It is extraordinary that this type of lesion appears so often on women during the Pre-Columbian period. They seem to have taken an active part in fighting (152). But such injuries may also have been the result of ritual ceremonies, as ethnology teaches us: West and Equatorial Africans, men and women, club each other when a tribesman is buried. If the lesions are multiple, as with the young man described by MacCurdy, this is mainly due to the fact that clubs were provided with a mace-head.

A cranial fracture of this type has been mentioned by E. Houzé (72) in an examination of the Neolithic men of Sclaigneaux, Belgium. The lesion may be regarded as having healed and has been described by the author as follows:

'Le pariétal droit présente en dedans de la bosse pariétale un enforcement de l'exocrâne de près d'un centimètre de profondeur, résultant d'un coup violent. La fracture a guéri; du côté de l'endocrâne il y a une saillie hyperostotique de réparation, présentant la forme d'un bouton aplati avec trace de fissure transversale.'

Apart from actual holes in the skull that have healed or not, a number of depressions have been described – varying sizes of indentations in the tabula externa, sometimes, but not always, with a similar effect on the tabula interna. These lesions have been put down to blows from axes and sling-stones, and in one case to an arrowhead. Such injuries are described by Le Baron on skulls from 'Caverne de l'Homme-mort' near Saint-Pierre-lès-Tripiez (Lozère), 'Caverne de

Lombrives' (Ariège) and from the dolmen d'Algérie, as well as on several skulls in the Broca Museum. The lesions are circular or oval with a diameter of about 7 mm. and in one case a maximum depth of 3 mm.

Many skulls show small bony outgrowths only about 3 mm. round and 1½ mm. long. These exostoses are best compared to drops of wax. They consist of very compact bony tissue and they easily flake off, leaving the normal bone tissue, without any impression, underneath. We must regard these exostoses as a result of traumatic periostitis.

In Iron Age skulls there occur oblong apertures in the frontal region, due to axe and sword blows.

The study of these traumata can provide interesting clues to the ways in which wars were waged in earlier times. For example a deadly sword stab behind the left clavicula in the cranio-caudal direction is known from the Roman period. Wells (187) has deduced from bone lesions the story of the glorious end of a mediaeval man found at Ipswich who may have been killed fighting on horseback. The skull shows a thin split 63 mm. long which runs diagonally across both parietals and has undoubtedly been caused by a blow from a sword. This injury can only have been inflicted by a warrior on horseback, unless the wounded victim had already fallen to his knees. Examination of the left femur provides the answer: a narrow cut on the antero-lateral side, slightly below the middle, indicates that a sword slash cut through the mass of muscles of the thigh. This is a typical wound inflicted on horsemen by footsoldiers, who always tried to keep on the left side of the horse's head, as can easily be seen from the examination of warriors' graves. As a result of loss of strength in the injured leg the victim can easily be pulled from his saddle. That this particular victim was a horseman is shown in the first place by a secondary wound on the front part of the left os ilium, exactly under the spina ilica ventralis, certainly cutting through the M. sartorius and the ligamentum inguinale; and also by the following three considerations: first, the small third facets of both distal epiphysis of the tibia, the result of the crouching or riding posture, secondly a healed fracture of the proximal extremity of the right fibula. This lesion is probably the result of a previous fight, because horsemen always tried to direct the right side of their horse towards the enemy, causing many such fractures of the fibula. Thirdly, from the presence of an exostosis of the dorsal face of the femur, referred to as a 'rider's bone', resulting from the tearing away of muscular tissue of the vastus lateralis from the M. quadriceps femoris and of the smaller head of the biceps femoris. Wells points out that once a rider has fallen or has been pulled from his horse he is extremely vulnerable. It may have been at that moment that this warrior received his head wound, though this cannot be known for certain. If a man is kneeling on the ground his back is most vulnerable, and injury of the vertebrae is likely. In this particular case the fifth cervical vertebra is missing. While this may be coincidence it is also possible that having been injured this vertebra was dissolved in the soil after burial. The blade of the right scapula suggests a lesion, though it is not really convincing.

The ribs too show evidence of fighting: the upper edge of the ninth rib and the lower edge of the eighth rib roughly in the middle of the bone on the right side show an incision made by a lance or dagger stab. The weapon was aimed at an angle of $45°$ and entered the thoracic cavity from behind and from above. This stab must certainly have pierced the right lung and the liver, and doubtless the loss of blood from the latter organ quickly brought on death. Two other lesions give us an idea of the knocking about this warrior had to endure: first a gash in the right shoulder which took away part of the acromion of the shoulderblade. This would have cut through not only the M. deltoides but also the ligamenta coraco-acromiale and acromio-claviculare which maintain the stability of the shoulder joint, damage which would almost certainly have eliminated the fighting ability of the right arm. Furthermore, on the same arm the distal end of the radius and ulna had been partly cut through. From these injuries it can be inferred that at that moment the man had his forearms bent and raised, with the humerus pointing downwards, a position which suggests either that he was going to strike, or that he was protecting his head against a threatening blow.

The study of traumata in palaeopathology is carried out on material which in fact provides only an indirect proof of bone injuries – that is, through callus around healed fractures, or inclusion of pieces of flint in the bone, which suggest open wounds. Displacement of bones in case of dislocation is not taken into account because excavation disturbs the anatomical context completely. Baudouin (11) however described a case of slipping of the atlas on the axis of a Neolithic skeleton. He established these facts during excavation, after having fixed the position of each bone in plaster immediately after it had been exposed. He was able to ascertain the following points:

1. All vertebrae—cervical, thoracic and lumbar – were undamaged, complete and in their precise anatomical position, with the subject on his back: except only the atlas.

2. The atlas was dislocated from the axis in such a way that the joint surfaces (which everywhere else were, without exception, in normal relationship) no longer articulated. That on the left of the atlas had slipped antero – medially from its counterpart on the axis. There was, in fact, lateral dislocation of the atlas, due to rotation forward and to the right, so that the odontoid process lay almost touching the posterior arch of the atlas. There was, however, no fracture of the dens. Doubtless death was instantaneous, as would ordinarily happen today in similar circumstances.

This case certainly shows how important it is to excavate the bones with as much care as the other archaeological pieces.

Fractures are recognized in growing numbers on Neanderthal, Cro-Magnon, Neolithic and Egyptian skeletons. The amount of material is too extensive for generalizations but nevertheless I should like to refer to a few characteristic cases.

First, exostosis of the femur of the *Pithecanthropus erectus Dubois* is assumed to be the result of a fracture – Baudouin (11) describes a badly healed fracture of the mandible as follows:

A consolidated fracture of the base of the left ramus, with overlapping of the bone, anteriorly and laterally, nearly as far as the first molar teeth; the second molar being almost completely hidden. Together with an extensive and irregular exostosis of the inferior border, from the symphysis to the second molar. The deformity is gross and typical. At the level of the apophysis geni two well marked hyperostoses, like large bony thorns, stick out horizontally and backwards.

Fractures of the vertebrae, resulting from staving in of the corpus vertebrae, may cause difficulties after healing. When two such vertebrae consolidate into one block it is possible that the pattern of spondylosis will be suggested; a radiological plate, however, will reveal the wedge-shaped vertebral lesions. I found a case of this type among the remains of bones from the cave of Antheit (67):

Among these latter were two vertebral bodies, fused to one another by a wide bridge of bone. At first sight one might suppose this was a spondylosis but radiography reveals a marked flattening of the height of the vertebral body, in its ventral part. From this we consider that it is more likely to be due to a fracture which has become wedge-shaped.

A characteristic example of traumatic lesion is the 'Man from Obourg', victim of the caving in of one of the flint mines exploited in the Neolithic period in the region of Spiennes (116). At the same time it is the first known work accident in Belgium (92) (Pl. 13). It seems worthwhile to point out one particular fracture of the ulna, ·

which throws a certain light on the way of living of certain groups of people. Neolithic peoples are usually taken to have been peaceful peasants. Surely the study of palaeopathology can prove the opposite. Studying the Neolithic race of Furfooz I found four ulnae which showed healed fractures (Pls. 14 and 15). The line of fracture occurred on the distal third of the diaphysis, twice on a right and twice on a left ulna. No corresponding fractures were found on the radii. However, we have to take into account two other facts: first, only fourteen radii were found for twenty-five ulnae. Secondly, in two cases the fracture was only partial, a so-called 'green stick fracture'. The line of fracture always occurred diagonally on the diaphysis, which proves a direct traumatic action, and, even more interesting, this traumatic factor always occurred on the ulnar side of the bone. Fractures of the ulna, especially by themselves, are rare. They certainly are so when they are indirectly caused by a fall, when the radius usually is fractured, as in the classical fracture of Pouteau-Colles. If sometimes the ulna is involved, it is usually the styloid process which is damaged. These considerations indicate that we are dealing with a special form of fracture, caused by the action of averting blows with raised arm during a fight. It deserves the eloquent name of 'parry fracture of the ulna'. I was able to identify this type of fracture on a late Roman skeleton excavated at Tongeren (fourth century A.D.). There are several references to these fractures (8). Houzé (72), writing about the Neolithic men of Sclaigneaux, describes a fracture of the proximal third of an ulna when no reduction took place, so that the two fragments form an obtuse angle and have joined by a huge callus. This callus had spread so much that the radius is enveloped in this excessive bony growth. Yet he mentions that he could not find any particular traumatic lesion of other bones which would point to ferocity of this people. My opinion on the other hand is that this lesion, together with the above mentioned fractured skull, are sufficient to prove the point. Rouillon (149) has described an even more eloquent case where not only callus is visible but also, apparently, signs of osteitis. This could easily mean a healed open fracture, which after being caused by a cutting weapon such as a sword or axe had developed a secondary infection.

Karl Sudhof (164) was surprised that 53·8 per cent of fractures showed complete anatomical healing. He considered that such a result could only be reached through orthopaedic intervention. Broca, De Nadaillac, De Mortillet, Lehmann-Nitsche, Le Baron and Vallois are also of this opinion.

The last named gives the following results for fractured bones:

| BONES | Number | Well healed | Badly healed | Doubtful |
|---|---|---|---|---|
| CLAVICLE | 4 | 4 | | |
| HUMERUS | 3 | 1 | 2 | |
| RADIUS | 12 | 12 | | |
| ULNA | 7 | 3 | 1* | 3 |
| FEMUR | 6 | 3 | 3 | |
| TIBIA | 6 | 3 | 1 | 2 |
| FIBULA | 3 | 1 | 1 | 1 |

[*pseudo-arthrosis]

Like Dr Raymond (144), I do not agree with this view.

A. H. Schultz (157) has proved that fractured bones of free living apes often heal spontaneously. Franjo Ivanicek (73) unknowingly confirms this opinion when he observes that mid diaphysial fractures of the upper and lower limbs healed well, but he specifies that this is not true of fractures where an extension is needed for a proper anatomical recovery, as in the case of a fracture of the femoral shaft. But still investigators subscribe to the sentimental speculation that prehistoric man must have had a good knowledge of orthopaedics. Le Baron (105) describes one single fractured humerus, where no shifting occurred during consolidation despite the sloping line of fracture. He thinks this successful healing would be possible only through the wearing of an orthopaedic apparatus. A few other cases that he mentions all point to healed but shortened limbs. Giot (60) marvels at the successful healing of broken ribs and a collar-bone with 'green stick fracture'! But next to this he describes a real fracture of the forearm, with less good result: the fragments of the radius fused in an angled position, so that the extremities of the ulna were unable to grow together and formed a pseudo-arthrosis. In their book (160) Shetelig and Falk, when talking about remains of bones from megalithic graves, state that: 'Broken bones occur, as is quite natural, and the fractured bone has often been very badly set.'

In the museum of Santander I have examined the burial contents of a child's grave from the Neolithic period, which had a femur with a fractured shaft, healed, but three centimetres shorter and with remarkable statistical change.

During the examination of the bones of human remains from the cave of Antheit, dating from the Iron Age (67), I had the opportunity to study a part of a diaphysis of the femur with stigmata of a healed fracture: the line of fracture is oblique – en biseau – running from proximo-ventral to dorso-distal. Both extremities of the bone had overlapped so that the bone had become shorter by about 4 cm.

The medullary canal is visible on each side but is divided by a ridge of bone which is part of the callus. The longitudinal axis of the diaphysis shows an angular deviation of about 10°.

Schmerling (155) describes a fracture of a fibula of a cave bear with *restitutio ad integrum* and says rightly: 'Une guérison d'une fracture transversale du péroné, sans recours de la chirurgie, ne surprendra personne'. Baudouin (11), speaking about a fracture of the lower third of the right tibia and fibula, which had healed completely in the correct position, also says: '. . . de telles fractures peuvent guérir parfaitement sans appareil. Et je suis bien certain que nombre de fractures modernes même immobilisées, ne sont pas mieux consolidées.' My opinion is that the main mistake lies in the fact that statistics are drawn up from all possible fractures, without any distinction between those types where spontaneous healing with almost complete anatomical recovery was possible and those which would have needed real surgical or orthopaedic intervention. Pales (135) however thinks it illogical to assume that prehistoric man should not have practised immobilization of a fractured bone, while monkeys, and even birds, do so instinctively. I agree with this, but immobilization itself does not ensure reduction or extension necessary for the proper healing of some fractures.

Reduction of broken bones which are enveloped in strong muscular tissue has, up to modern times, been a difficult task, powerful muscular spasm being an important hindering factor. Relaxation of the muscle is necessary and this is only possible with proper anaesthetics. Reduction of a dislocated shoulder without anaesthetics is always dangerous because of the risk of fracture. Anaesthetics are of extreme importance not only as a pain killer, but even more to achieve that relaxed condition which until recently was quite impossible.

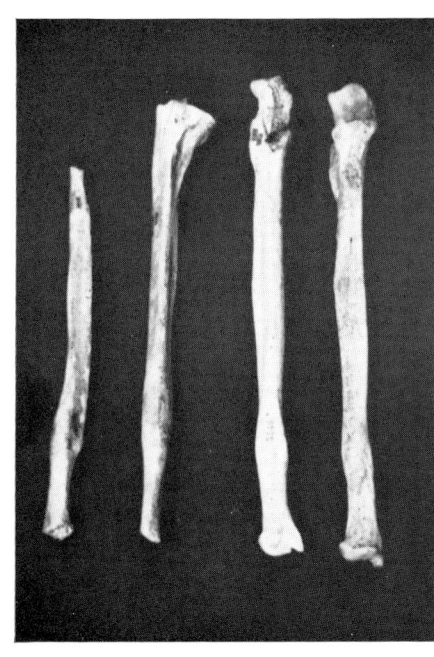

14. Four ulnae with well-consolidated fractures. Race of Furfooz.

15. (below) Radiographs of the same bones.

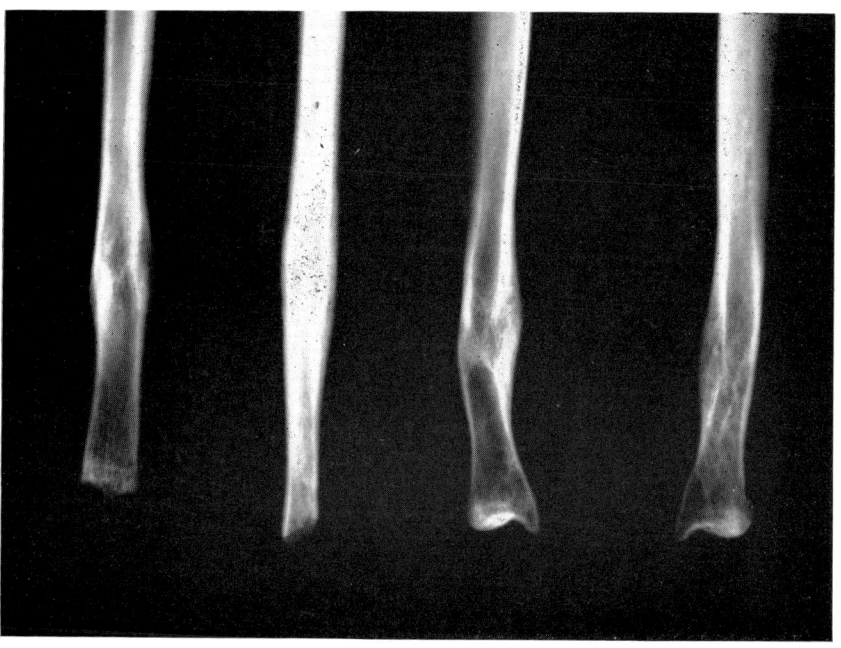

16. Human vertebra, in which is lodged a Neolithic arrowhead. From M. Boule, *Les Hommes Fossiles*. Cave of Lozère.

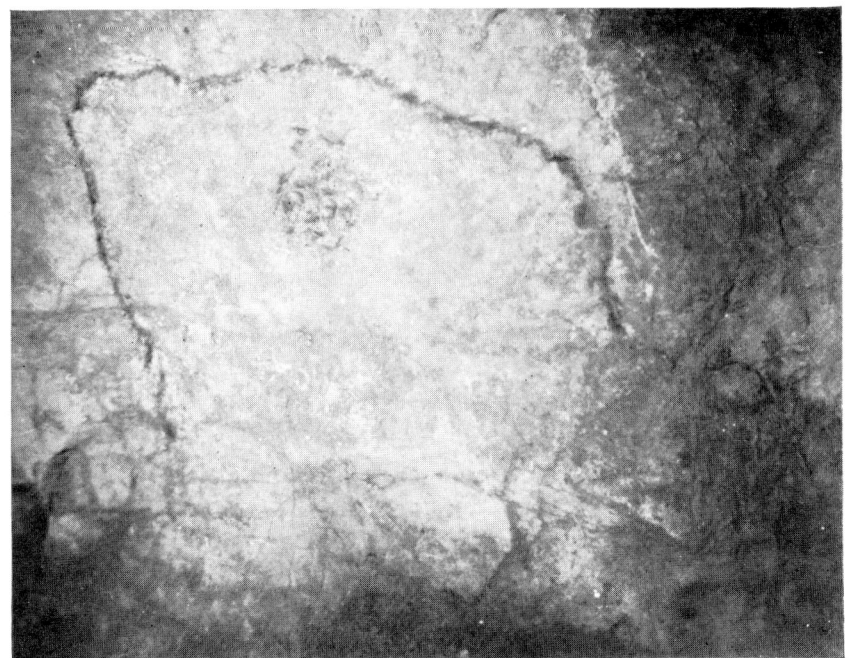

17. *Elephas primigenius.* Cave of Pindal (Spain).

18. Deer. Cave of Covalanas (Spain).

D

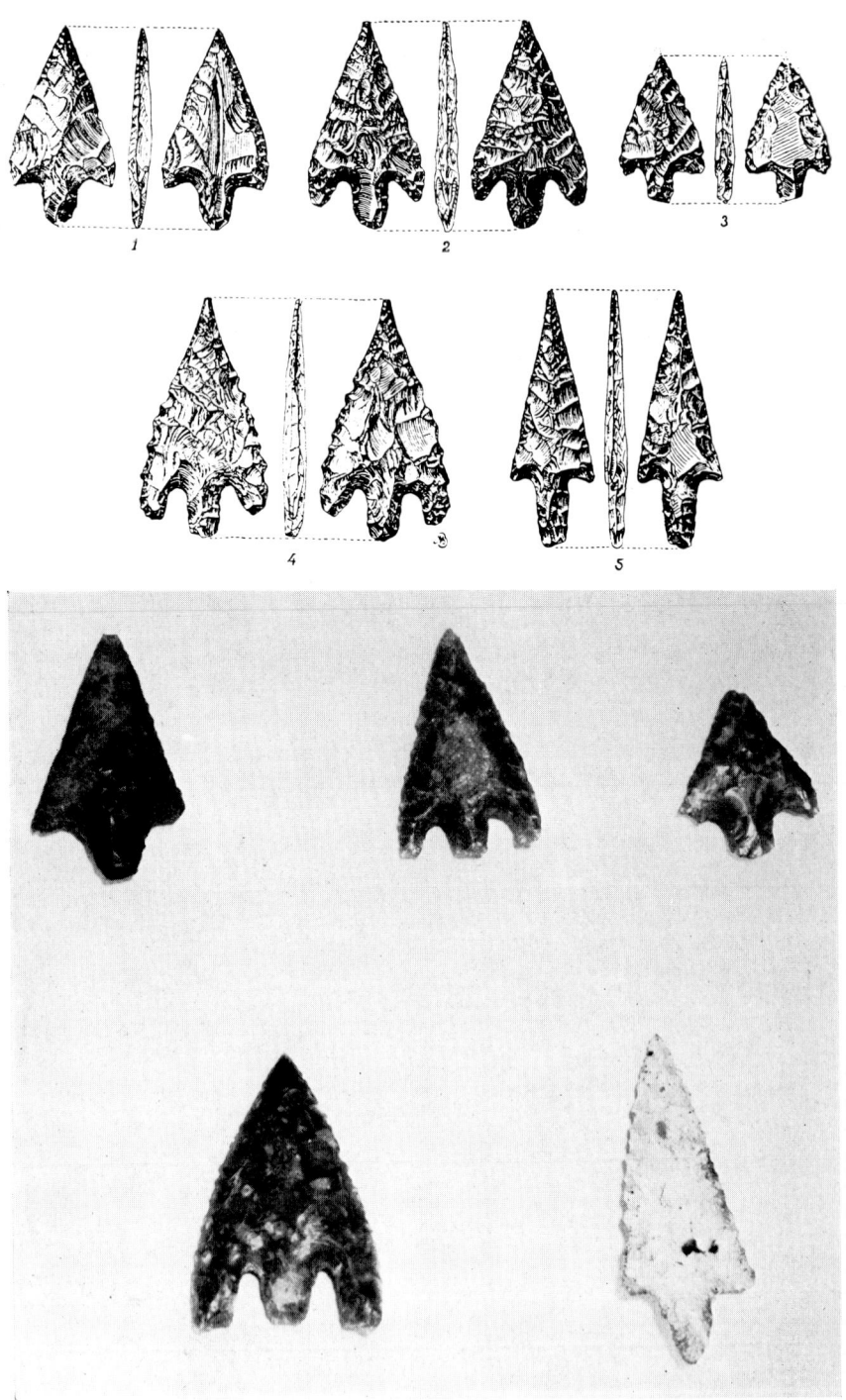

19. (top) Arrowheads from the upper Solutrean culture, cave of Parpallo. From Pericot Garcia, *La Cueva del Parpallo*.

20. Similar arrowheads found in the Belgian Kempen, dating from the Neolithic period. Left to right: (1) Lommel Fabrieken. (2) Lommel Dorperheide. (3) Deurne Brem. (4) Overpelt Houtmolenheide. (5) Overpelt Houtmolenheide.

CHAPTER 9

# Injuries from stone weapons

Injuries from pointed flint implements have been detected in great quantities not only in animals but also in humans. Pales (135) mentions a number of bones from the Palaeolithic in which the fatal piece of flint was found, but none of these was human. In Toulouse there is a vertebra of a Pleistocene deer recovered from a Magdalenian layer in the cave of Montfort (Ariège), in which a pointed bit of flint is lodged. The whole vertebral body was pierced and part of the flint penetrated the medullary canal. A. Wouters (194) found a typical Gravettian point in a mandible of *Cervus giganteus* recovered from Dutch Limburg.

C. Wells (191) relates the case of a Tardenoisian triangle, embedded in the sixth thoracic vertebrae of one of the Mesolithic skeletons of Téviec (Brittany), but to find real pedunculated arrowheads one has to wait until the Neolithic. Linear injuries discovered on skulls are often taken for lesions caused by arrowheads, but such linear grooves do not necessarily have a traumatic origin. Wells (189) describes them on tibiae of both men and women. He regards them as normal vascular impressions and sometimes made by the anastomoses of the periosteal veins which may enlarge with osteitis. These can also occur with arteriosclerosis when they follow a twisting course. Le Baron (105) describes a cranial vault from the *allée couverte* at Equihen near Boulogne-sur-mer, with a bow-shaped lesion 22 mm. long, the upper part of which perforates the cranium in such a way that a piece of bone has been chipped off the tabula interna. There are no signs of recovery and the lesion is situated in the sulcus of the meningeal artery, so that bleeding may be regarded as the cause of death. However, few skulls are found with the arrowhead still lodged in them. I should perhaps mention the skull from Aurignac in which a piece of flint was found lodged in the

D

left temporal squama (55). Morel (127) describes a child's skull with an arrowhead lodged in the temporal bone. This piece can be seen in the museum of Oran.

The situation is different with other bones. Sometimes, when a fragment of the arrowhead is still lodged in the bone, this shows that an attempt has been undertaken to remove the injurious object. In other instances, such as wounds of the ilium, the arrowhead has been discovered in the bone and covered with healing, reactional tissue, which indicates total recovery. The same can be said of a left ulna dating from the Neolithic period, which I had the opportunity of studying in Toulouse Museum. It shows a barbed arrowhead which had not been removed and which had penetrated the bone along the facies radialis. The point had entered so deeply that only the tang and the two little barbs were protruding. But already the space between the point and barbs had grown over with bony tissue, again indicating a complete recovery.

A great number of vertebrae have been found pierced by flint points (Pl. 16). Sometimes the injury healed, at other times it led to a quick death, depending mainly upon the direction in which the arrowhead entered the body. The anatomical relation between spine and aorta is here the important factor. It should be mentioned that this type of lesion has also been found in cases where the arrowhead concerned was of the transverse type (7). These lesions give us an idea of the efficiency of Neolithic flint arrowheads and again prove that these people, although usually described as peaceful peasants, did fight and kill one another, as I have already mentioned when discussing the bone lesions found on skeletal remains from the cave of Furfooz.

Flint arrowheads can even be identified as the injuring or killing agent in cremation burials. Beex (14) discovered a triangular Neolithic arrowhead with a flat base among the cremation remains of a grave which probably dates from the Bronze Age. The flint is badly affected through fire, though more so at the base than at the point; the end is broken off. We are perhaps justified in assuming that the arrowhead caused death and, further, that on entering the tip broke off in the bone, in such a way, however, that the part penetrating the flesh was more protected against the heat of the fire than the protruding base (Pl. 9).

This discussion of whether prehistoric man had any notion of orthopaedics, and the fact that he certainly tried to remove arrowheads from the body bring us to the further question of whether he had any notion of anatomy.

Sigerist (161) speaking about the embalmers in Ancient Egypt says 'They had some anatomical knowledge to be sure, the same kind of knowledge that the butchers and cooks had, or the priests who sacrificed animals to the gods, but no more.'

Anatomy is a science which has known a slow evolution, with ups and downs. Even dissections on living criminals resulted in faulty interpretations. Galen mentions these practices and says that Herophilos and Erasistratus, who both lived in the time of Alexander the Great, carried them out. Their writings were lost when the library of Alexandria was destroyed by fire. Notwithstanding these facts prehistoric man is often regarded as having had a rudimentary knowledge of anatomy. It is my intention to see whether any facts can be inferred from the study of the many pictures left us by prehistoric artists, including engravings as well as paintings, and on both cave-walls and utility articles.

# CHAPTER 10

# Prehistoric art

Prehistoric art has only come into its own as a chapter in the history of culture during the last few decades. Andrews (3) relates that not thirty years have elapsed since Hooton observed that the enthusiasm of students of prehistory had caused them to lose their sense of values so that they had begun to overestimate these attempts.

Due mainly to the industry of eminent art historians prehistoric works of art are now considered things of beauty. They move us in just the same way as the most famous paintings in an art gallery.

Cave paintings were discovered accidentally. Their recognition was rather dramatic. In 1875 the little daughter of the Spanish Marquis Marcelino de Sautuola attracted her father's attention to the paintings she saw on the ceiling of the cave of Altamira (province of Santander, Spain). The Marquis was excavating there after seeing prehistoric tools in the world exhibition at Paris. The cave of Altamira had been discovered some years before by a hunter whose dog had chased a fox into a freed split in the entrance which had been blocked for centuries. The Marquis's discovery was dismissed completely and he was even accused of falsification, as he had been entertaining a French painter as his guest. It was twenty years before prehistoric art was given recognition.

In 1895 Rivière discovered drawings in the cave of La Mouthe. These appeared after scraping layers of glacial period: here was proof that the pictures were at least as old as the cultural layers which had covered them. Those doing research in prehistory realized their mistake and justice was done to the Marquis de Sautuola, although this late recognition came after his death. So far more than a hundred prehistoric caves with wall paintings have come to light.

Prehistoric art first appears in the Aurignacian period. What is remarkable is that these first paintings are neither primitive nor stylized but, on the contrary, rather naturalistic and impressionistic.

Exceptions are the magical sexual figures, which at a very early date were already completely abstract (see below). It should be pointed out however that prehistoric art before all else is magical art. A second important point is that prehistoric artists have managed to master the problem of reproducing masses, that is to say reproduction of a group which the eye is unable to distinguish in detail. Only one or two figures are accentuated, the others are not sharply outlined. In our time only the impressionists have mastered this technique (103).

In the Aurignacian culture the figure is reproduced realistically and in a sensuous manner. Only the outline is important without depth. One front leg and one hind leg are reproduced in such a way that the picture is two-dimensional. Discovery of the third dimension was only possible with the use of colour or when thickening of the line breaks through the plane level. Polychrome was first employed in the Magdalenian, a beautiful example being the frescoes of Altamira.

We can discern a transition from the Aurignacian linear technique to the Solutrean pictorial technique, which was perfected during the Magdalenian culture. Animals are depicted running, jumping, dying or giving birth, in a polychrome of merging colour planes, and like impressionism the style gives truthful representation of a momentary sensation. At the end of the Magdalenian period, however, linear techniques occur again, but different from those used in the Aurignacian: the lines no longer define, but rather, as H. Kuhn says, 'in what is at the same time a plastic and pictorial way they give expression to the essence of the represented subject'. In this he sees a relationship between prehistoric art and modern expressionism (103). This form of art is only to be found in the caves of eastern Spain: the human figure, which seems to have been taboo in the art from the glacial era, comes to the fore. Again it is not the figure which interests the artist, but much more its movements. This expressionism of abstracted forms develops into the symbolism found in the Andalusian art. The painting material consisted of mixtures of fat and red or yellow ochre, black manganese and iron oxide; products which were readily available in nature.

Up to now no prehistoric cave paintings have been found in Belgium. Traces of ochre on cave walls however indicate the possibility of their existence in earlier periods. The damp climate might have been the cause of their disappearance. Yet beautifully engraved figures on bone or stone have been found. They were recovered from caves which had been inhabited by prehistoric men in the glacial period. These objects are displayed in the Natural History Museum in Brussels.

The picture of the well-known but badly faded Aurignacian mammoth figure from the cave of Pindal (Spain) (Pl. 17), is regarded as the first known anatomical drawing. A blot on the figure is usually assumed to be the animal's heart. In his work about mammoths Maska (119) toys with the idea that this blot might be nothing more than a very wide elephant's ear. He takes into account, though, that ears of that size are not characteristic of the mammoth. An analogous picture of a stag, with a similar blot, has been found in the cave of Covalanas at Ramales (Spain) (Pl. 18). The drawing consists of lines, which themselves join points. As this technique is considered to be the oldest, we can assume that this picture is a lot older than the one from Pindal. It is my opinion that in the first case the artist has tried to colour the whole figure, and that in the second, the stag, the artist made a mistake when trying to draw the front legs, which are missing in any case.

As these drawings are believed to have an exclusively magical character connected with hunting, the alleged 'heart pictures figures' are interpreted as marking the 'vulnerable spot'. This certainly is not right, for particularly with larger mammals the heart is not the most vulnerable spot.

It is certain that flint spear-points from the Upper and Middle Palaeolithic period could not penetrate the body very easily. Spears made exclusively of wood, with a cut point hardened in fire must have been much more effective. It is not certain whether one can justly infer from cave figures that the bow and arrow were known in the Palaeolithic age. It is true one sees feathered arrows or spears in paintings in the cave of Lascaux. But these are from a later phase. The bay leaf shape, from the Solutrean, looks more like an arrow-head, and even more do the barbed flint arrowheads of the Upper Solutrean from the cave of Parpallo (Spain) (137, 138) (Pl. 19), whose shape is not much different from one found in the Belgian Kempen, which dates from the Neolithic period (Pl. 20). We may assume that the effectiveness of piercing weapons was raised by the use of poison. Grooves in bone and horn arrowheads and spearpoints may be an indication of its use (Pl. 24). Rust (152) is convinced that bow and arrow were used. Indeed in Ahrensburg near Hamburg he found a hundred wooden arrows and two fragments of a bow, made of pine which had been preserved in layers of peat. They date from the last phase of the Palaeolithic period (Pls. 21 and 22). He distinguished three types of arrow and concluded that their use must have been known for a long time, even in the Mousterian. That these were indeed arrows to be used in hunting is shown by the finding of the skeleton of a wolf which had had its vertebrae pierced by an arrowhead which was found in the spinal canal. Rust also found many shoulderblades of reindeers with a cavity which could only have been caused by arrowheads.

Yet on the other hand, finds in Moravia point mainly to the use of pitfalls (110). Other finds, such as those in Solutré, point to the use of natural precipices, over which the game was driven. In the cave of Lascaux one can see the picture of a horse, which plunges downwards with raised front legs. In the cave of Castillo (Spain) a bison is pictured in red ochre with its head down and hindlegs up; the animal's longitudinal axis is vertical. In the immediate vicinity of the cave the meadows end in a slope which narrows and ends on a rock, behind which gapes the precipice. The resemblance with Lascaux is striking. In the same Spanish cave a natural rock protuberance has been worked into a bison's head. Right underneath on the wall is a small black *signe tectiforme*: it looks as if the animal is being driven towards the pitfall. Tectiform figures are linear pictures on cave walls, whose meaning has so far eluded us, though they are often associated with the use of pitfalls. It is to be noticed that these from Castillo are accompanied by a series of dotted lines. These dots are also to be found in La Pasiega (Spain) it is clear here that the lines curve around a quadrangular figure. In this figure the picture of an animal can be faintly discerned. Perhaps prehistoric man represented water or a river by such a dotted line. In other words the quadrangular figure would be a pitfall placed along a river, at a spot regularly frequented by animals for drinking (Pls. 23, 24, 25).

Almost certainly the primitive attached little importance to the heart. He could see how beasts of prey killed their victims through biting them in the neck, where opening of the jugular veins causes a quick death. The primitive himself chooses the belly as the most vulnerable spot: indeed the skin is very thin there and easy to pierce. The horse in Castillo and the bisons of Niaux and Pindal illustrate this eloquently. In the unique hunting scene in the cave of Lascaux a human figure is shown in front of a snorting bison, whose intestines bulge out of the abdomen (Pl. 26).

In my opinion one particular engraving on a *baguette décorée* is to be counted quite definitely among what we regard as anatomical drawings (76). The piece is on display in the museum at Les Eyzies. It was found in the cave La Madeleine (municipality of Tursac) and dates from the Magdalenian culture (Pl. 27). The picture represents a more or less bent penis. The glans penis shows distinctly the mouth of the urethra and is surrounded by a widely protruding preputium, which suggests a circumcized phimosis. In the middle of the corpus cavernosum a medial line is noticeable, which starts at the corona glandis and ends near the anus. This line may be interpreted as the urethra. Since on the one hand there is a gap in the opening and, on the other, this line is situated in the extension of the orificium, so that the drawing shows clearly the direction of the splitforming which is anatomically right – we can also assume that this is meant to be the corpus cavernosum urethrae. The hindmost part shows the raphe perinei, and two lateral parallel lines trace out the perineum.

On both sides of the base of the corpus there are two oval bodies, which certainly represent both testes. These are shown dissected. The two strokes in front of them seem to me to represent the scrotum, rather than the epididymis. Bladder, prostate and bulbus urethrae have been left out. In front of the penis is an engraving of an animal's head.

I believe that another example may be identified in De Laet and Glasbergen (33). In the chapter about Aurignacian culture both authors describe an object as '. . . a piece of reindeer horn, engraved with mysterious signs'. This piece of horn still shows the general shape of the part of the antler from which it was made. It is 100 mm. long, and was found by E. Dupont in the Trou Magritte (Belgium) in an early Perigordian layer (Pl. 28). F. Twiesselmann (158) has studied this piece thoroughly and has compiled the opinions of different authors about it. E. Dupont thinks the piece deserves no special interest; E. Van Overloop considers that the engraving is meant only to be decorative and has been adapted to the overall shape of the piece; A. Rust sees it as the picture of a swan; C. Ausselet-Lambrechts, H. Breuil and R. de Saint Périer regard the figures as two spools, shaped as fishes, showing close resemblance to Palaeolithic pictures from Central Europe. According to Twiesselmann himself it might be an ideogram or a schematic plan of a cave. None of these suppositions seems satisfactory to me. My opinion is that the two figures represent the sexual organs. The lower one represents the uterus, which is continued into the vagina. The latter is correctly curved round the pubis. Quite possibly too the vulva is represented. The top figure shows a penis. The whole was probably a magical representation. This engraving, executed in the very earliest phase of glacial era art, already shows certain signs of abstraction at a moment when the art is still far from its culminating point.

The same abstraction is noticeable in a 40 mm. high 'Venus' figure in mammoth ivory found by Dupont in the same layer.

Breuer (17) says of it: 'Le sens des signes gravés ne paraît pas avoir jamais été bien défini.' Again it is clear that products of prehistoric art have a magical meaning whether they are two- or three-dimensional. This magical meaning is especially apparent in the engraved piece of reindeer horn, and abstraction is in fact nothing more than putting the emphasis on an important detail, a *pars pro toto*, which in due course becomes a symbol.

I discovered a similar abstract engraving on a piece of bone 90 mm. long, from a bird, which is part of the collection of R. Robert. The piece dates from the final phase of the Magdalenian culture. The figure (Pl. 29) has been described as a fish by L. R. Nougier (131):

The engraved bone represents a fish. Its outline is firmly and regularly drawn, with its maximum bluntness in front and sharply narrowed at the tail. The line of the vertebrae or the lateral line organ of the flank is well rendered by a straight stroke. Two rows of fine, parallel hatchings, grouped in pairs, indicate the scales. The caudal fin, which is the only one shown, is made of five groups of small lines which depict its concave shape. Some lines unfortunately truncated by an ancient breaking of the bone, occur in front of the engraved fish and may perhaps represent the tail fin of another one swimming ahead of it.

I cannot agree with this account because Stone Age artists represented all animals realistically, and the result is a very clear figure, precise in every detail (Pl. 30). In the picture on this piece the tail is not in proportion to the body, furthermore the eye, the mouth, as well as the ventral and dorsal fins are missing. The median line is too central to represent the spine. Again my opinion is that it represents the uterus, with the median line representing the uterine cavity. The muscles are shown by diagonal dashes, while the tail represents the vagina with the vaginal rugosity. The organ is represented apart, as if it were dissected.

I have found two other descriptions where authors have taken the picture to represent the female genitals. The first is of a prehistoric pendant of which A. de Paniagua says (13):

In the cave of Tuc d'Audubert, there was found a pendant amulet, pierced for suspension, which is nothing but a vulva. It is of antler, 40 mm. long, and has the form of an elongated triangle with an incurved base. It is perforated above by a round hole, about 8 mm. across, from which a deep gutter descends across one side of it, widening towards its centre and ending at the base of the object. The whole surface is scored with short lines, going from right to left and perhaps representing hairs (this opinion is debateable). The gutter is the cleft between the buttocks, prolonged across the anus to the vulva. On the other face, which is un-worked, the suspensory hole is surrounded by a kind of rim – the large vulval lips.

The second is again a piece of R. Robert's (147):

The other object is also the upper end of a perforated baton, with most of the hole, the two sides of which show deep radiating incisions. There is no question of a vulva here unless it is a symbolic vulva encircled with hair, such as are known from other examples.

The foregoing pictures illustrate what I should like to call 'ana-tomical figures'. They may be only the reproduction of details, but even if they have a magical meaning and are already quite abstract, they are still valuable in that they are pictures which teach us some-thing about prehistoric man's knowledge of anatomy. A further point of interest is that these abstractions already occur in the Aurignacian, whereas they are usually assumed to occur only from the Azilian culture onward (132).

CHAPTER 11

# Prehistory –
# tradition and folklore

I have already pointed out that in front of the penis on the *baguette décorée* of Tursac is engraved the head of an animal. The occurrence of this anatomical picture together with an animal figure points to the magical character of this decorated object. 'Man's compelling action on nature is the origin of magic', says Maxwell. Broca says: 'Among all peoples, magic before being based on observation, draws its origin from superstition.' The same can be said of medicine, and the two notions are closely related. Even more interesting is the fact that medical magic outlives its real practitioners. Flint tools continued to exist for a long time, for qualitative as well as ritual reasons. At the beginning of the metal age flint served many more purposes than bronze and even iron, which at the outset were weak. Graves in the Iron Age contain more stone than iron for magical reasons. In Brittany 'men-gurum', or 'thunderstones', which are no more than polished axes, are believed to prevent lightning from striking (109). In Lorraine people thought that the same *pierres de tonnerre* fell from the sky to a depth of seven feet in the earth, slowly to come to the surface again in another seven years (172). They were even believed to have specific healing powers: a warmed axe placed underneath a cow's udder and sprinkled with a few drops of milk, which had to be allowed to evaporate, was thought to prevent the udder from swelling. Worn against the kidneys it was believed to be a remedy against renal calculus: in 1600 the Count of Lorraine was presented with such an axe to fight this disease.

Not only the axe, but also those who use it, the woodcutter or the carpenter for instance, are credited with magical power. For instance they are believed to be able to heal phlegmons of the hand. In Spain these polished axes are called *piedras de rayo* or 'lightning stones'. A flint knife can be used to the same purpose. In Belgium (130) they were sometimes cemented in a wall when a new house was being built. In Brittany until this century a dying person could ask in the presence of the fellow villagers to have his skull smashed with a polished stone axe. This custom was built within the framework of a Christian rite. The stone axe was called *le marteau béni*. Zacharie Le Rouzic describes such a stone on display in the Miln Museum. Could there be any relation between this custom and the tradition of tapping a deceased pope three times on the head with a silver hammer?

If in a community an object acquires a magical meaning, it will be cherished by the community. The axe, the most important tool, is such an object. In Schleswig-Holstein, until recently, an axe was placed on the threshold of the cowshed when the cattle left it, a practice thought to give protection against disaster. As a magical tool an axe does not require a complete socket, which is difficult to make anyway: that is the reason why the socket of so many axes is incomplete. Axes too small to be used as an ordinary axe, are probably magical implements; usually they are now called votive axes. A reduced symbol for the axe is the St Andrew cross (58). When the inventory of the Dom of Utrecht was drawn up in the thirteenth century, a so-called 'St Martins hammer' was discovered. It is shaped like a prehistoric hammer-axe, mounted in silver and provided with a wooden handle. Legend has it that St Boniface was killed with this axe.

The use in Egypt of flint knives for embalming has been discussed in the introduction. In Belgium there is a lime tree in which 70,000 nails have been hammered to exorcize evil spirits (120). This is a ritual custom passed down from earlier generations that still lives on in a region rich in Neolithic finds. An oak used for similar purposes is known in Herchies (Hainault, Belgium) (52). Nails are hammered into its trunk to obtain recovery from furunculosis, and there are Walloons who speak of *clous* when referring to furuncles. Again it is striking that Herchies is situated in the vicinity of Spiennes, which is known for its Neolithic flint mines.

# CHAPTER 12

# Disturbances in development

The development from fertilized ovum to fully grown individual is a relatively slow and complicated process in mammals. Embryology teaches us that nature is very conservative and that evolutionary stages follow the pattern of already existing animals. The human embryo passes through all existing forms between single celled organisms and mammal. This complicated process should progress along an absolutely harmonious course. The least defect may result in unforeseeable and drastic deformations in the future individual. Teratology, the study of monstrosities, investigates the extremes of this sort of defect.

Congenital disturbances may result from causes within the cell nucleus itself or from adverse factors affecting the foetus during its intra-uterine development. In the first group we are concerned with genetic factors such as mongolism. In the second group several factors may cause deformation: for example, infections such as congenital syphilis, toxic cases such as the notorious thalidomide, and other factors which are still not sufficiently understood as in the case of harelip and cleft palate.

Of course the deformation is not always so terrible, and in many cases the defect may be negligible or not apparent at all, not only for other people but even for the person who is afflicted with the lesion. Neither is it necessarily the skeleton which shows defects, since soft-tissue organs can also be affected with lesions, for example when the septum atriarum fails to close. Of course the former are of primary importance to palaeopathology, though many belong even more to the field of anthropology. I mean for example metopic suture, which divides the frontal bone in two halves by continuing the sagittal suture from the bregma to the nasion; wormian bones; scaphocephaly, a narrowing of the cranial roof, which takes the shape of a keel as a result of the premature fusion of the sagittal

47

suture; or the deviations of the distal epiphysis of the humerus known as perforatio olecrani and epicondyle projection. These abnormalities are more important to establish the degree of kinship between individuals, who for example, have been found in the same burial place. A good example of this is referred to by Wells (191), where an Anglo-Saxon group from Gloucestershire showed several cases with bifid ribs. Deformities can also occur during lifetime, as a result of endocrine disturbances often affecting the skeleton – we refer to a number of cases, which have been compiled mostly by Pales (135) and Sigerist (161).

In the first place there is the perforated sternum of a Neolithic man (135). This lesion was already present in the embryo, as the sternum originally consists of two cartilage halves. The opening may be very small, but can also take the form of a real fissura sterni congenita, when the presence of the foramen causes the sternum to be widened. The opening always occurs in the corpus and is situated centrally. With South American Indians it seems to occur in 13·3 per cent of all cases, with Europeans in 6·9 per cent (117). MacCurdy has found it in about 10·7 per cent of all cases in Pre-Columbian Peruvians (113). Baudouin has described two cases from Vaudrest (11).

This lesion has been recognized on the well-known skeleton of Avennes (Belgium), from the Neolithic period (39, 81) (Pl. 31). Although Pales has described this deviation as commonplace, I could not find it in any of the Neolithic skeletons from the province of Namur (72). I should like to stress the fact that perforations of the sternum may be the result of the erosion caused by aortic aneurism. If so it is apparent that the lesion is the result of real osseous necrosis due to pressure, and that its position is in no way connected with the metamerization, which means that the lesion is not necessarily situated centrally and that also the manubrium may be perforated (118). Because of the syphilitic aetiology of aortic aneurism it is absolutely impossible to find such lesions in Belgian prehistory (see chapter on syphilis).

Next there is the cranial malformation, attributed to hydrocephalus, of a mummy from the Roman period – a man of about thirty years old. The condition of the brain caused partial hemiplegia of the left side of the body with inevitable atrophy of the bones of that side (D. E. Derry). Derry, D. E. (1913) *J. Anat. Physiol.*, 47: 436–59.

Further, an open sacrum from the Neolithic period is known in France. Latent spina bifida often occurs in grown-ups, while spina bifida usually causes early death in babies (135). Still Gerhardt (57) has described a typical case of spina bifida in a young woman (fourth century A.D.) of the first six cervical vertebrae: the dorsal part of the arcus together with the processus spinalis are missing. At the same time he found identical lesions in three sacral vertebrae. MacCurdy (113) describes several cases of open sacrum, sacralization and lumbarization of the fifth lumbar vertebra and of the first sacral vertebra in Pre-Columbian Peruvians. Raymond (144) mentions a case of congenital dislocation of the right hip.

Talipes equinovarus occurred occasionally. Siptah, Egyptian king of the XIXth dynasty, suffered from this defect. Smith and later also Ruffer regard it as a congenital malformation. Dr Slomann, however, points out the possibility of poliomyelitis, contracted during puberty, as the cause; indeed the lesion is that of a pes equinus without the accompanying varus deviation, which would have been established had the lesion been congenital (Pl. 32).

Dwarfs have also been found from Ancient Egypt. Indeed they were popular then, at least achondroplastics were because dwarfism of this type is not combined with lack of intelligence. Well proportioned dwarfs who do not show any lack of psychic functions, are rare. Nam-Hotep, from one of the last dynasties of Pharaohs was an achondroplastic. His statue was found in a necropolis at Sakhara (161). The skull of a cretin dating from the XVIIth dynasty is also known.

A femur from Lozère was believed by Pales (135) to show the characteristics of Paget's disease. This disease, also called osteitis deformans, occurs twice as frequently in men as in women. Mainly skull and long bones are affected, unlike osteitis fibrosa cystica generalisata, or von Recklinghausen's disease which affects all bones. Paget's disease is characterized on the one hand by osteoporosis, and on the other by new bone formation. A radiographic picture of the affected bones shows a thickened cortex with lamellar construction. Osteoporosis gives the pelvis a cloudy look and the skull is affected in the same way (23). The femur described by Pales (Pl. 33) shows a diaphysis, with very irregular thickening. The edges are rough and the bone's pronounced convex curve is striking. The shaft of the bone gives a blown up impression and looks like mouldered wood. If one cuts the bone, the inside shows rarefaction of the compact tissue, so that it looks like network, with anastomosis at the very fine osseous lamellae. In 1912 Professor Poncet diagnosed Paget's disease on the skull of a mummified Cynocephalus from Ancient Egypt. He based his diagnosis on the symmetrical hypertrophy of the cranial bones: projection of the frontal and suborbital zones and relative thickening of the cranial vault. Both facts point to Paget's disease. Yet there is no porosity nor the 'cloudy picture' with unclear outline. On the contrary, the outline of the tabula externa is extremely clear. Pales thinks it may be a case of 'goundou', a form of tropical framboesia. The femur in question had also been described, before Pales, as affected with syphilis. Notwithstanding the clear radiographic picture, I want to add that a bone affected by gummateous osteomyelitis never bends but always breaks, in contrast with Paget's disease, where the bone does bend.

Two humeri found in the dolmen of Meudon in 1869 are described by Pales, with due reservation, as being affected with von Recklinghausen's disease. Both bones show a thickening of the diaphysis and especially of the metaphysis, and collapse of the humeral heads. Radiologically they show a decrease in the thickness of the compact tissue to that of a scale. The marrow canal is filled with fine osseous lamellae, which indicate a fibrous transformation of the marrow. It is true, no real cysts can be indicated: the small cavities spread through the top part of the bone hardly deserve this name. In any case there is bilateral fibrous osteitis.

Schlaginhaufen has described a femur from the Neolithic period, found in Switzerland, which he believed to have belonged to someone suffering from acromegaly. This is a hasty conclusion, for it is the skull which shows the most characteristic signs of acromegaly, particularly the sella turcica, because of the abnormal pituitary gland. Thus Sir Auckland Peddes describes the very thick cranial roof of the former Piltdown specimen as affected by acromegaly, while Moodie thinks it is affected by Paget's disease (124).

Senile osteoporosis should not be regarded as a separate pattern of disease, although it is useful for establishing the age of a skeleton or a mummy. 'Old age is not a disease and cannot be treated', writes Minot. 'It is an accumulation of changes, starting at youth and continuing during the whole life of an individual.' Osteoporosis however is not necessarily linked to senility. It may occur in young people. An osteoporotic skull shows porous zones, blown out with many tiny cavities which make it look like pumice. When a slight osteophytic reaction takes place, it has the appearance of foam or sponge. Typical of the lesion is that it is bilateral and symmetrical (Pl. 34). The disease affects particularly the parietal bones. Yet the lesion can also be found on the os frontale, occipitale or temporale; and rarely on the sphenoid. It is too characteristic to be confounded with passages dug by the beetles post mortem. Hrdlička discovered

21. Two extremities of bows in pine wood. From Rust, *Die Jüngere Altsteinzeit*. Excavated at Stellmoor.

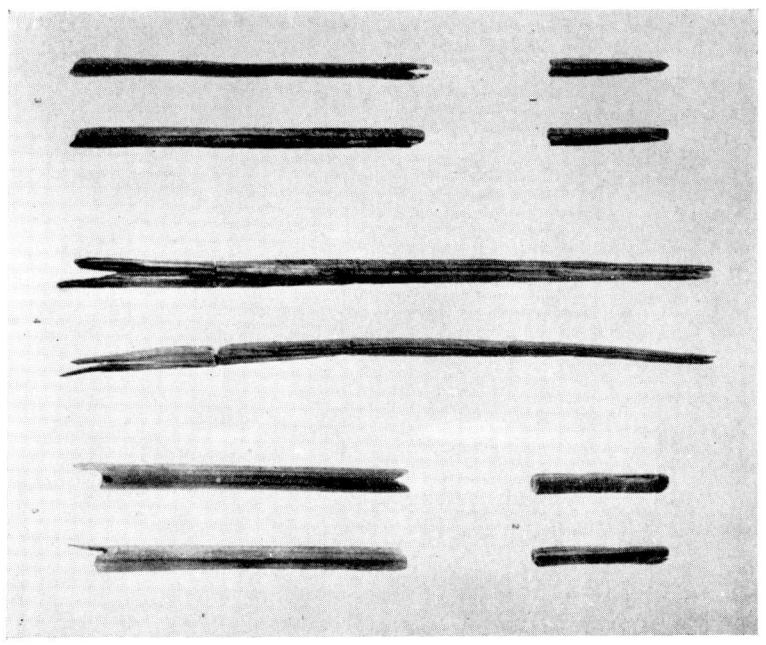

22. The same excavation. Wooden shafts of arrows. Notice the bilateral notches for bow string and fixing of the flint point.

23. Pitfall, partly surrounded by series of dots (river?). In the middle the outline of an animal is vaguely to be seen. Cave La Pasiega (Monte Castillo)

24. Tectiform figures, with series of dots (river?). Cave of Castillo.

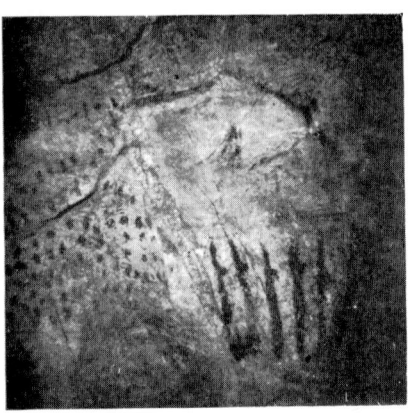

25. Bison, wounded in the belly by stabbing weapon. In front, a series of claviform signs (humans?). The dots in front of the animal probably represent water of a drinking spot. Cave Pindal.

three cases of this disease on the long bones of Pre-Columbian children. The disease seems to affect young people most readily. If the lesion is on the exocranium, no change will be noted in the endocranium. Sometimes the lesions spread to such a degree that the compact tissue is unrecognizable and the skull sutures fuse. The fact that zones of compact tissue sometimes occur in osteoporotic zones is taken to be an indication of recovery. Radiographically the lesion looks like orange-peel in front. At the side the skull looks as if it is planted with short, very closely standing hairs ('Hair-on-end' appearance). Histologically the diploë looks lacunar. There is a certain similarity with the pattern of rickets.

Up to now, no such lesions have been found on Palaeolithic men. One single case of a child is known from the Mesolithic period where the lesions are very superficial. Elliot Smith found symmetrical depressions of the parietal bones on Egyptians from the upper class. He ascribes them in part to the custom of wearing heavy wigs, and sometimes he regards them as congenital or senile atrophy. In Nubia similar lesions have also been found.

Pre-Columbians in particular are affected. The distribution of the lesion here is also strange: in North America the lesion seems uncommon, among the Mayas it occurs frequently, and even more in Pre-Columbian Peru. In the latter region only coastal people are affected, not mountain dwellers.

Parrot raises the possibility of syphilis. Others have considered rickets. This disease however did not occur among Pre-Columbian Indians, though it does exist in Peru now, while Peruvian skulls today are never affected by osteoporosis. The same applies to Egypt. Wood Jones ascribes the lesions to carrying heavy loads on the head. This is not well grounded however, first of all because it is mostly children who are affected by osteoporosis, and further because the practice of carrying loads in this way today does not result in this lesion. Moreover, this theory fails to explain the localization around the orbits.

Williams and Adachi devote their attention solely to the possibility of disturbance of the blood circulation. This might be the case with new-born babies, when intentional deformation of the skull causes pressure on the occipital bone, which would result in venous stasis. This moulding of the skull was still practised in France until recently, but did not cause particular osteoporotic lesions: and although deformation was a slow process no serious lesions appeared. This, however, is again in contradiction to cases in America, where cranial deformation was also practised and was the cause of brain

51

E

damage. Moreover, the American Indians put their children in a hanging cradle with the head lower than the legs, which increases stasis. The absence of osteoporosis in adults may be due to the fact that osteoporosis occurring in a child may later heal. Osteoporosis caused by cranial mutilation could be connected with meningioma, which can arise from irritation of the meninges, indeed, there is a certain radiological similarity between both conditions. Hanseman does not agree with this view, since he found analogous lesions on skulls of monkeys, where there is no artificial irritation of the brain.

Moodie regards osteoporosis of this sort as a result of nourishment disturbance due to infection of the dura mater, which starts in youth.

Osteoporosis, also called cribra cranii, is usually associated with cribra orbitalia. These are small holes in the roof of the eye sockets, connected with the paranasal sinuses, but not with the brain (124).

Besides senile osteoporosis and bilateral cranial osteoporosis there also exists a pre-senile form, which affects mainly the spine (95).

The disease consists of a progressive melting away of bone. Not only is there insufficient calcification, as with osteomalacia but also the spongy bone tissue disappears. In both cases deformation may arise, though based on different mechanisms: osteomalacia softens the bone, osteoporosis causes a greater brittleness of the bone.

An aetiological relation between osteoporosis and insufficiency of the sexual glands may exist. Osteoporosis may occur in women after the menopause, but young men also may suffer from the disease, which is usually attributed to the action of certain steroid hormones of the adrenal gland in inhibiting the working of the osteoblasts.

The difference between pre-senile osteoporosis and senile osteoporosis is that the first form attacks the spine, while the second is diffused over the whole skeleton.

Differential diagnosis should take into account osteoporosis of the spine in endocrine diseases such as Cushing's disease. Osteoporosis of the whole skeleton can occur with hyperthyroidism, hyperparathyroidism, hypogonadism, and diabetes, as well as with certain liver affections.

The difference between osteoporosis and osteomalacia has been mentioned already. Inanition-osteopathy is in fact a stage between the two. The plasmocytome can show a similar pattern to osteoporosis. Furthermore von Recklinghausen's disease, early carcinoma and tuberculosis of the spine will cause certain difficulties – although to a lesser extent for separate osseous remains because of their demineralizing and necrotic effect.

A typical case of lumbar osteochondritis has been described by C. Wells on the lumbar vertebrae III and IV of a girl of about sixteen, from the Bronze Age (*c.* 1600 b.c.) found in Dorset (England) (184).

Alkaptonuria or ochronosis is an affection which could be described as both a genetic and a metabolic disturbance. An article has been written about it by Wells (186). It is shown by the urine turning dark brown to black, shortly after micturition. It is a very rare disease with an occurrence of about one in ten million people. The only clinical symptom is that the presence of homogentisic acid, which causes the dark colouring of the urine, causes certain changes in the intervertebral discs; these become brittle and are seen as white strips on a radiograph, through condensation. The chance discovery of two cases in mummies caused Wells to conclude that the disease must have occurred more in the past than now. Further cases now lead him to think that this appearance is an artifact, in some way due to the process of embalming.

As the study material at the disposal of palaeopathology is relatively scarce we should make use of all possible sources. Several disturbances are known from pictures, which I shall refer to more extensively below. I have already mentioned that developmental disturbances are easily recognized; for instance achondroplasia, which is to be found in skeletons and in mummies, as well as in pictures. Also worth mentioning is a stele in the Louvre dating from the year IX of king Sesostris I. The sculpture is of an important official, Hor, son of Sent-Mait, XIIth dynasty. The male figure has a large breast with nipple clearly visible near the left armpit. On the right side there is no trace of such a development, so that we may wonder whether this is a case of unilateral gynecomastia.

An unidentified lesion of the right kneecap has been described by Baudouin (12). The bone is big and heavy and shows signs of osseous rarefaction. There is a gap in the shape of a crescent on the leg, 20 mm. long and 15 mm. wide. Baudouin rules out fracture and also an abnormal ossification because the patella has only one ossification centre. Yet he takes into account the possibility of a supplementary ossification centre which has not united with the first one. This anomaly, of which several variants exist, has been further mentioned by R. Martin (117).

Morel (127) describes a number of vertebral deformations which he regards as hereditary, found in one group of Neolithics from Roquefort. The lesions differ. Some consist of a fusion of vertebrae to form one block, in such a way that spondylarthrosis is ruled out.

The blood supply of this block consists of one artery for both vertebrae (lumbar vertebrae one and two). The spinous process and the transverse process have each melted to a separate mass. Another case of classical lumbarization of the first sacral vertebra is known. Moreover, Morel describes a vertebra which besides having characteristics of the fifth lumbar vertebra also has characteristics of the first sacral vertebra. As the first dominates, he believes it is a case of lumbarization rather than of sacralization.

Last he describes a number of vertebrae which are asymmetrical: the vertebral bodies have deviated from their axis sideways. At the same time the vertebral arch and the epiphyses are asymmetrical as well.

Rouillon (149) describes an atlas in which an osseous canal had formed around the arteria vertebralis. His description is as follows:

The first vertebra, instead of having a *sulcus* on the superior surface of the posterior arch, behind the glenoid articular facets, for the passage of the vertebral artery, shows on the right side a true bony canal with roof, which is very thin moreover, and reduced almost to a bony thread. On the left this thread is represented by two *spicula*, anterior and posterior, which do not meet, so that on this side the canal is incomplete.

Baudouin (11) describes a case of change in the anatomical form of an atlas of a woman, as a result of its new function in an abnormal atlas-epistropheus joint:

The malformation, which occurs on the right side, affects more particularly the right superior articular facet. This is reduced to its posterior half. The anterior half is wholly absent and is replaced by a deep depression, triangular at its base, provided with a large vascular foramen. The anterior part of the habitually elliptical or kidney-shaped cavity (which is normal on the left, although divided in two by a tiny transverse crest) is absent therefore on one side. As a result of this, the right articular surface, displaced backwards, is almost circular. It measures $12 \times 14$ mm. in contrast to the left facet which is $9 \times 20$ mm. From this, it very closely resembles the free surface of the left inferior epistrophic articular facet. What is more, the right inferior articular facet, situated beneath it, is altered. Instead of being circular, which is usual, it has become elliptical and measures $19 \times 10$ mm., whilst its fellow is $14 \times 14$ mm. It has the shape of a small female facet. It seems, therefore, that it is the anomaly of the superior facet – due to a developmental arrest of the anterior part of the articular cartilage of the right glenoid cavity – which has caused the change in shape of the inferior articular process on the same side. Because the right occipital condyle cannot slide from front to back, in the absence of a right glenoid cavity, a new articulation has had to be produced on that side between the occiput and the atlas. This demonstration of articular plasticity strikes me as very remarkable. Once again it shows that it is surely function which moulds organs. What is most extraordinary, is that the transformation has taken place on only one side at the level of the atlanto-epistrophic articulation. It follows that, when the person bent her head to the left, the movement was between occiput and atlas and that, on inclining to the right, it was between atlas and axis. On the other hand, when she raised her head the whole movement could take place at the atlanto-occipital joints.

MacCurdy (113) describes a so-called 'node of Kerckering' in a Pre-Columbian skull from Peru, where the occipital bone has failed to develop in the region of the opisthion.


METABOLIC DISTURBANCES

Some endocrine disturbances cause such characteristic deviations that these are easily recognizable by sight. Prehistoric reproductions could be useful here, but portrayals of humans are quite rare in these periods. Possibly prehistoric man felt the same fear for a picture of a man as his primitive brother today. If man is pictured, the figure is usually grotesque or masked and in many cases the features are very sporadically indicated. Only in the cave of Vienne (France) do we find figures with more or less clear features.

In her excellent book about human figures, E. Della Santa (154) gives a survey based on the different ways in which they were reproduced. In the first group she distinguishes those pictures of humans with a recognizable animal's head, including those of birds, mammoth, horse, cow, goat and deer. Altogether there are twenty-seven reproductions of this sort. A second group of thirty reproductions consists of creatures with an undefined animal's head. The third group comprises grotesque characters with disproportionate features. She found about forty-six of these. With this third group she discusses bas-reliefs and figurines, together totaling forty-one specimens. In a fourth group she brought together realistic pictures, totalling only thirty-six.

# CHAPTER 13

# 'Venus' figurines

From this title it is already clear that the human figure from Palaeo-lithic times was not only reproduced in drawings. It was cut in stone or carved in ivory. These sculptures consist almost exclusively of female figures, which are often called 'Venus' figurines. They all originate from the Upper Palaeolithic period, in particular the Old Perigordian or the Lower Aurignaceo-Perigordian.

These figurines were made between 20,000 and 40,000 B.C. In Belgium one not very beautiful specimen, carved in mammoth ivory, was found in the Trou Magritte (Prov. Namen) (167).

Art of this sort once again confronts us with the mind of pre-historic man, who wonders about life and the universe, which he does not understand, though he is aware of its secret powers, which seem friendly at one time and hostile at another. We can understand that he will try to protect and defend himself against these powers.

G. H. Luquet (111) divides the 'Venus' figurines into four groups: an academic type, which gives a picture true to nature; a steato-pygous type, which is only recognizable as such in profile; a steatonic type, in which fat accumulations on hips and thighs are noticeable both in profile and in front view; and a fourth type in which all parts of the body are conspicuous through their obese degeneration. Absolon differentiates seven types or groups, based not only on the presence of pronounced sexual features, but also on the presence or absence of facial features and other non-sexual physical qualities (140).

Generally most striking in the 'Venus' figurines is the exaggerated emphasis on the secondary sexual characteristics, the so-called steatopygia, which is also a characteristic of female Bushmen today. Certain writers dispute steatopygia in a few figurines, which they prefer to call only adipose. Dr Verneau (176) regards the figurines from Brassempouy and Lespugue, as well as the bas-relief from Laussel, as steatopygous; Dr Radmilli (141) agrees with him. Sigerist

(161), however, regards the figurines as representations of the 'Great Mother', source of life. Della Santa (154) puts forward five possibilities: first, the figurines could be a real picture, true to nature; this we can accept for some forms but not all, certainly not those which have strongly exaggerated secondary sexual characteristics. For this reason Hoernes and Menghin and also Klaatsch (98) think that these figurines fulfil the aesthetic ideal of women in prehistory. Klaatsch finds an explanation in the change of diet: Neanderthals followed a more strict vegetarian diet than the Aurignacians. The latter fed mainly on meat. This diet, it is argued, would have promoted the desire to create objects with erotic significance. Della Santa rejects this theory because eroticism as an artistic ideal is seldom found in hunting peoples. C. Schuchardt thinks they might be figures of priestesses, basing his theory on two bas-reliefs from Laussel on which the female figurine holds an object which looks like a horn. He had already opened another field of vision, assuming that they could be representations of ancestors. Hancar (154) agrees with this, since figurines were found in a niche during excavations in Russia and Siberia. We could assume some sort of veneration, which can easily concur with Sigerist's supposition. The latest interpretation of the nature of the 'Venus' figurines is that they might be fertility symbols. Among the supporters of this view are Count Bégouen, Piette, Boule, Hoernes, Menghin, Goury (63) and even Luquet, although the last has certain reservations.

J. A. Manduit (120) goes so far as to regard the figurines as fertility symbols for animals. His argument is as follows:

Through inability to grasp the difference between objective facts and his own subjective impressions everything about him, i.e. prehistoric man, was real; mental pictures were identical with reality and partook of the same attributes.

On the one hand, hunting magic at this remote epoch when hunting provided the basis of man's subsistence, enabled him to supply the essentials for survival and, on the other, eroticism, which assured life's continuance, equally acquired great importance. These two vital needs, so interwoven in primitive thought that certain peoples, even today, have not differentiated them in their language and still use the same word to express both love and hunger, were the basis of magic practices. By a similar chain of reasoning, prehistoric man concluded that human eroticism, which took care of the increase of mankind, ought to act in the same way on the fecundity of animals.

Luquet has also thought of this possibility. I cannot agree with this view, knowing that H. Kuhn (102) puts Genesis to the test as it were, and situates the Earthly Paradise in the glacial era, when there was plenty of game in the form of vast herds of horses, bison and reindeer. With the exception of this last argument, usually nothing is said about the purpose of these figurines. I should like to assume that they are symbols for the maintenance and increase of the race, although Luquet does not see any gain from multiplying the race where there was likely to be continuous fear of shortage of food (111).

Prehistoric man, paradoxical as it may seem, must have led a very intense life. This intensity was aimed at satisfying the inborn urges, for food and propagation. Food was gained through hunting, and finds point mainly to big mammals as the spoils of the hunt. It is important to realize that the primitive weapons did not allow a man to kill a big animal single handed. Pitfalls were necessary (see chapter on traumata), and this method of hunting called for hunters to chase the game in planned and organized fashion. These conditions can only be fulfilled with a fairly large number of men. Maintenance and even increase of the race was an absolute necessity: the individual on his own was lost and helpless in the last glacial era, especially when he had to hunt mammoth and bison. I believe that the complete lack of carved human figures from the glacial era in Spanish pre-history may confirm this view. Indeed, climatic conditions were much better in Spain and fauna consisted mainly of horse, stag and deer, which are vulnerable to bow and arrow. Only one 'Venus' figurine has been found in Spain, in the cave of El Pendo (Santander), and even that one is in my opinion a very dubious specimen, for it looks more like a broken *bâton de commandement* or a rough trial. (Pl. 37).

Several investigators have ventured on an estimate of population density in prehistoric times. De Laet (32) thinks that in Belgium during the last glacial era about 400 individuals should be assumed for one generation, which means a density of one person per 75 square kilometres. He reached this number through comparison with the population density in Canada in the nineteenth century, where the tundra regions may have provided living conditions quite similar to those in Belgian regions during the last glacial phase. Leroy-Gourhan (108) has worked out on a basis of food requirements a density of one prehistoric man per 30 square kilometres assuming that at most five reindeer can live on one square kilometre of tundra. L. R. Nougier (133) supposes that in Gallia during the last glacial era there lived about 50,000 individuals. This density did not change

26. The well-known hunting scene from the cave of Lascaux.

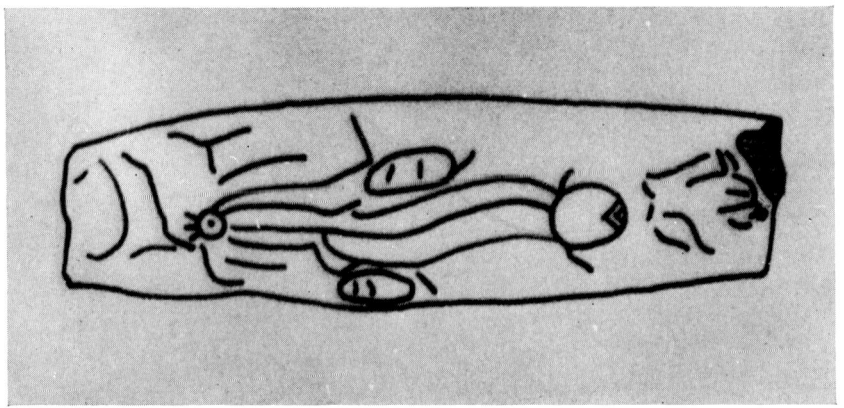

27. Drawing of a *baguette décorée*, Magdalenian culture V, found at Tursac (France).

28. Decorated object in reindeer horn. After F. Twiesselmann.

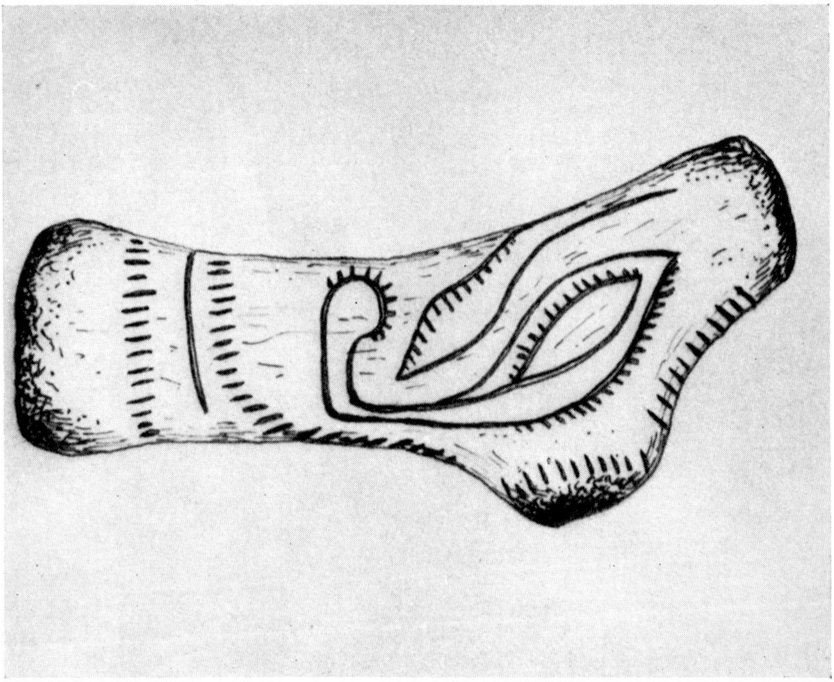

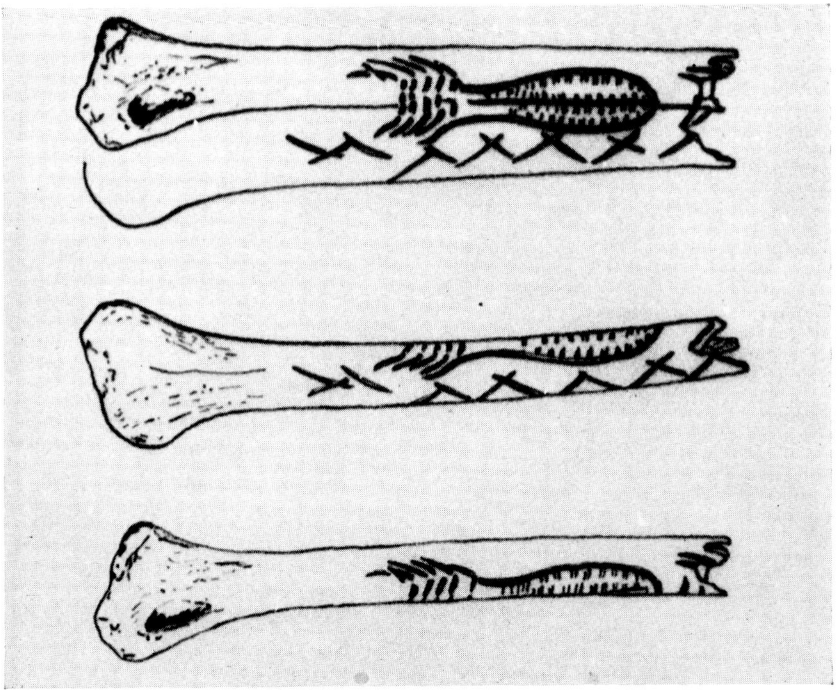

29. Decorated bird bone. Collection R. Robert.

30. Deer and fishes engraved on a piece of bone. Museum at Les Eyzies.

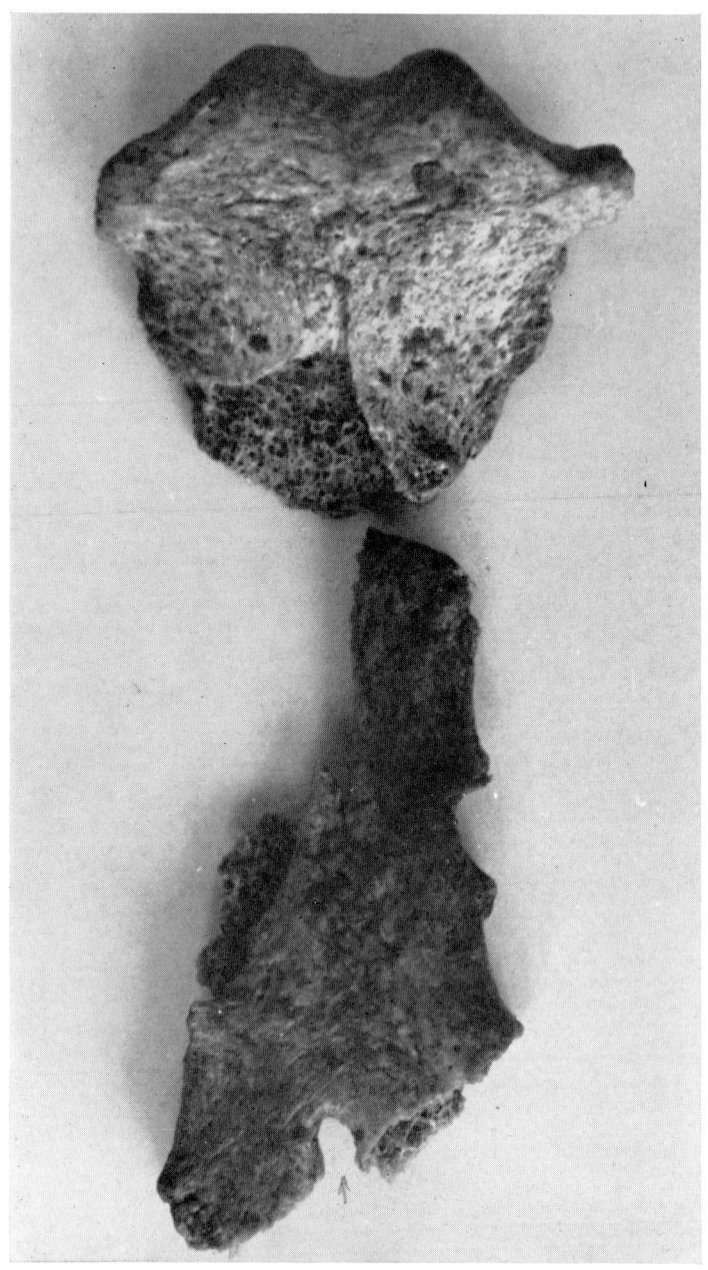

31. Part of the sternum of a male skeleton found at Avennes by Destexhe-Jamotte. The line of fracture runs through the perforation.

during the following Mesolithic period. Only from the Neolithic period onward is it assumed that population density must have increased sharply. Nougier estimates that a total population of 10,500,000 individuals lived between 2600 and 1200 B.C. This means 20,000 for each generation, assuming the average life was twenty-five years.

The 'Bandkeramik' settlement of Köln-Lindenthal in the initial phase of the Neolithic period is reckoned to have comprised 250 to 350 inhabitants. Goury (63) brings forward yet another considera-tion, namely that the greatest number of figurines are found to belong to the Aurignacian culture, which is characterized by quite a mild climate, while they become fewer and fewer in the Solutrean and Magdalenian cultures, when the climate grows colder. This he explains on the one hand by improvement in soundness and effec-tiveness of weapons, and on the other hand by the increasing cold which might endanger food supply, so that surplus population was to be avoided. Luquet even claims that the killing of new born babies in times of over-population and famine is a rule with some hunting peoples. Another factor is that the advent of the new race, Cro-Magnon man, must have caused 'clans' to arise. These 'clans' could only keep their ground through numerical strength. To me therefore an important factor for the interpretation of the 'Venus' figurines seems to be the average duration of life of prehistoric man.

# CHAPTER 14

# Average duration of life

Investigation of most skeletal finds teaches us that most of the people died round about fourteen years of age, in other words when they were still children. Procreation possibilities of a girl of fifteen are quite small, though being forced to live together in caves may have prompted promiscuity among primitives. Maclaren (114) says it is a very common thing among Australian primitives, although he does not stop to consider why so few young women get pregnant. Some investigators explain this as follows: they assume that because of daily sexual intercourse, the spermatozoa in the testes do not get the chance to develop completely. True, Maclaren does say that when fertilization takes place and the child is born alive, mother and child are killed out of fear of inbreeding. To account for taking the significance of the 'Venuses' to be fertility symbols we have to consider two factors: on the one hand a tribe with many members, and on the other, the reduced procreation possibilities of these members, mainly due to their young age.

If the average duration of life increases with progress in civilization, we still note a great difference between Vallois' figures for the Neanderthals and the figures for the burial mound of St Urnel-en-Plomeur. Vallois (168, 169) gives the following figures for mortality before the twentieth year: 55 per cent for Neanderthals, 34·3 for Cro-Magnons, 37 per cent for Neolithics. He provides the following mortality figures for Neanderthals compared to the figures for Austria in 1927:

| Age | Neanderthals (per cent) | Austrians (per cent) |
|:---:|:---:|:---:|
| 0–14 | 40 | 15·4 |
| 14–20 | 15 | 2·7 |
| 21–40 | 40 | 11·9 |
| 40+ | 5 | 22·6 |
| 60+ | — | 37·4 |

The number of Neanderthal skeletons found to date however, does not allow us to accept these figures generally. Indeed Vallois based his calculations on a total of seventeen individuals:

*Age*

| | |
|---|---|
| 0–11 | 5 |
| 12–20 | 3 |
| 21–30 | 5 (1 man, 4 women) |
| 31–40 | 3 (all men) |
| 41–50 | 1 (1 man) |

Besides these another seventeen individuals are known whose ages have not been established, though it is reported that eight were very young, five were young adults, and two were older but not senile. He has also established the following figures for *Homo sapiens fossilis* of which fifty-three individuals are known in Europe from the Upper-Palaeolithic period:

| | |
|---|---|
| *Children* | 15 |
| *Adolescents* | 7 |
| *Adults* 21–30 | 13 (4 men, 6 women, 3 unidentified) |
| 31–40 | 11 (5 men, 5 women, 1 unidentified) |
| 41–50 | 6 (5 men, 1 unidentified) |
| 51–60 | 1 (1 man) |

Besides these there are five more adults whose ages could not be established but who were certainly not senile. For the equivalent period in Africa (Afalou-Bou-Rhummel: Ibero-Maurusian period) he gives the following figures for forty-five individuals:

| | |
|---|---|
| *Children* | 6 |
| *Adolescents* | 3 |
| *Adults* 20–25 | 11 (4 men, 7 women) |
| 25–35 | 11 (4 men, 7 women) |
| 35–40 | 10 (8 men, 2 women) |
| 40–50 | 4 (3 men, 1 woman) |

For the Mesolithic period in Europe he has combined the figures of skeletal finds from Ofnet, Téviec, Hoëdic, Montardit, and Gramat, totalling fifty-eight individuals:

| | |
|---|---|
| *Children* | 18 |
| *Adolescents* | 2 |
| *Adults* 21–30 | 29 (12 men, 17 women) |
| 31–40 | 6 (4 men, 2 women) |
| 41–50 | 1 (1 man) |
| 50+ | 2 (2 men) |

For the Natufian period in Erg-el-Ahmar, Palestine, he gives:

| | |
|---|---|
| *Children* | 1 |
| *Adolescents* | 2 |
| *Adults* 20–30 | 2 women |
| 30+ | 1 man (not senile) |

We can see from these figures that of 173 individuals only three were older than fifty years.

Krogman (101) values the following comparison:

*Neanderthals*    80 per cent were dead by the age of 30 years.
                 95 per cent by the age of 40 years
*Cro-Magnons*    61·7 per cent were dead by the age of 30 years
                 88·2 per cent by the age of 40 years
*Mesolithics*    86·3 per cent were dead by the age of 30 years
                 95·5 per cent by the age of 40 years

For Sinanthropus he gives the following figures from twenty-two skulls:

| *Age* | |
|---|---|
| 0–14 | 15 |
| 14–30 | 3 |
| 30–50 | 3 |
| 50+ | 1 |

Which means that 68 per cent were dead by the age of fourteen years. A point arising from these figures is that mortality in young women is higher than in men. Seldom does the primitive survive until the physiological decline of the secreting glands, which means that the menopause or the andropause occur seldom in primitives. Civilization was responsible for this change through antibiotics and opotherapy.

Mortality among young people must be even higher than would seem from the above figures, for, in my view, children's skeletons are less well preserved in the soil. This fact is very well illustrated by the figures for the cemetery of St Urnel-en-Plomeur (Bretagne) where the particular soil conditions had caused the skeletons of babies and children to be preserved as well as those of adults. The burial place dates from the Iron Age; the top layers were identified as La Tène I and II, the bottom as Hallstatt. 225 skeletons were recovered (60):

50 per cent babies and very young children
25 per cent 6–20 years (mainly 6 and 7 years)
25 per cent adults of which: 15 per cent were between 20 and 30 years
                             5 per cent were between 30 and 40 years
                             5 per cent were over 40 years.

Only one individual was older than sixty years: the skeleton showed a calcified thyroid cartilage, as well as ossification of the manubrium of the sternum. The mortality figure indicates that 75 per cent of the population died before the age of twenty, compared with 55 per cent among the more primitive Neanderthals: a conclusion which has no justification.

# CHAPTER 15

# Deficiency diseases

RICKETS

When we consider the primitive way of life of Palaeolithic man and even that of the more recent Neolithic man, involuntarily deficiency diseases come to mind. We can imagine that the initially quite severe climate, even if it alternated with less cold periods, and hostile nature around him, against which man was insufficiently armed, must often have endangered food supply. I have already referred to magical drawings which sought to ensure rich hunting. Many investigators accept that cannibalism must have been common, not so much for ritual reasons as from need. The smashed *Sinanthropus* skulls may indicate cannibalism, as well as the skulls from the cave of Ofnet, and those which show an artificial enlargement of the opening in the back of the head (19). Yet these facts do not seem to agree with reality, as we have already seen in the discussion of bone lesions. The most we should assume is that an insufficient and one-sided diet, almost exclusively consisting of meat, may have had its side-effects.

Deficiency can reveal itself in the bones of Palaeolithic men in several ways: the most notable are avitaminosis D and avitaminosis C. The latter we know as scurvy. The former occurs in two forms according to the age of the sufferer, namely rickets in young persons, and osteomalacia in adults. At first sight it seems that rickets must have occurred frequently in the Pleistocene period. There are a number of connected factors however, the most important being the quality of the food. Avitaminosis D, or rickets, does not occur in primitive peoples such as the Eskimos, who have a diet rich in cod liver oil. Similarly Palaeolithic man does not seem to have suffered from rickets thanks probably to his carnivorous diet. Hrdlička (135) has found no signs of rickets in Pre-Columbian America: this is probably due to intense sunrays, because the ultra-violet rays convert the ergosterol in the skin into vitamin D. The same applies to Ancient Egypt: Wood-Jones examined 6,000 skeletons in vain, no trace of the disease was found. Furthermore, child mortality was high and burial places always contain many skeletons of children, which should increase the chances of finding rickets. Yet occasional cases do occur, as is clear from representations of figures with bowed legs from the grave of Beni-Hassan. A mummified baboon also showed these symptoms. But running about naked in the sun was the best preventative against this deficiency disease. In Palaeolithic times the eating of raw food must have been a further factor, as vitamin D is not destroyed in raw food as it is after cooking (170).

Rickets does occur in the Neolithic period; it was found in bones from Norway and Denmark. The hunter has become farmer. Civilization takes a step forward when the hunter polishes stone and makes compound tools. He was forced to do this anyway as the fauna had changed: the big mammoth and reindeer herds had gone north, with the retreating ice-barrier. Man depended more and more on the food which the soil provided, and tried to speed up its production. Even if his ignorance of manure and the resulting exhaustion of the soil forced him to search for other land, his wanderings would have been less extensive than those of a hunter who had to follow the game, which is not so much tied to land as to a particular kind of food. Agriculture requires hard labour for which man's powers are not sufficient or at least are poorly rewarded. Domestication of animals, in particular of the horse and cow, was an important step, and perhaps we can say that surgical intervention in the form of castration of the bull was the most important step in this direction (161). The domestication of sheep and invention of weaving techniques were also of great importance. Clothing more effectively shut out the diminishing cold though the body was now deprived to no small extent of the sun's rays. This contrasts sharply with the Palaeolithic period in which clothing must have been minimal, as Ivanicek (73) concludes from his finds in the Slav necropolis at Ptuju, where many arthritic characteristics were found on vertebrae and joints. He agrees however that these deviations do not necessarily imply arthritis but may be the result of specific traumata and infections. His observations agree well with those, already mentioned, made by Desse and Giot (38) in connection with late-Neolithic finds. The slight clothing worn in cold climates attracted the attention of Darwin, who noted that in Tierra del Fuego children ran about naked in biting cold weather and that adults dressed only in a single guanaco skin (125).

We see from this that civilization has promoted rickets (166). Rickets is the disease of dark, slummy town quarters in the North – not so much because of the cold but because of lack of sunlight and a meat diet. The disease occurs only from the Hallstatt Iron Age period onward (eighth century B.C.) (74), although cases have been described of bones affected with rickets from megalithic graves in Denmark (160). Schmerling (155) describes a certain number of bones of cave bears, affected by the disease. In his description of the lesions he uses general terms such as 'caries', 'necrosis', 'exostosis' and 'degeneration' and he thinks that outward causes are at the basis of these lesions. The many fistulas in the lesions together with cavities and bone reactions surely suggest rather a chronic osteomyelitis; moreover at the time Schmerling described the lesions, the pathological pattern of rickets was not yet clearly established.

In adults symptoms of deficiency of vitamin D lead to osteomalacia. Decalcification of the bones and the ensuing malformations usually start during pregnancy. The bones can become as pliable as rubber. This may not be rickets proper, for certain unknown factors are involved. Moodie (124) even mentions a case of osteomalacia in a primitive carnivorous animal from the Eocene, but this diagnosis is not certain. The animal, *Lymnocyon potens*, which lived 3,000,000 years ago in Washakia (Wyoming, USA) shows decayed hyperplastic lesions on the lower part of the tibia and fibula, the tibio-tarsal joint and some of the tarsal bones.

SCURVY

Avitaminosis C, or scurvy, can cause bone lesions, but only during the process of growing. These occur on the growth cartilage, and are mainly of vascular origin. Although this disease should be more apparent in a carnivorous diet than the often absent rickets, the available material does not allow us to detect the disease, as is clear from previous data.

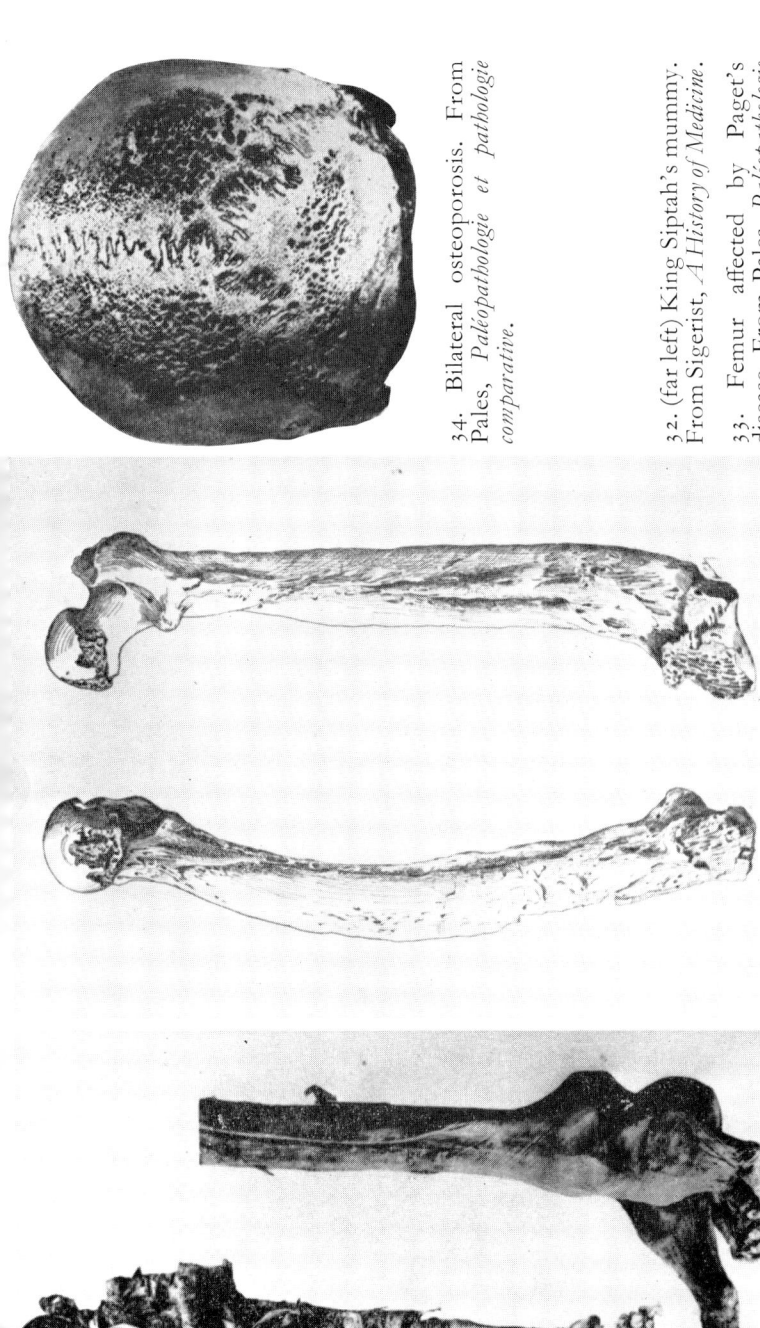

34. Bilateral osteoporosis. From Pales, *Paléopathologie et pathologie comparative*.

32. (far left) King Siptah's mummy. From Sigerist, *A History of Medicine*.

33. Femur affected by Paget's disease. From Pales, *Paléopathologie et pathologie comparative*.

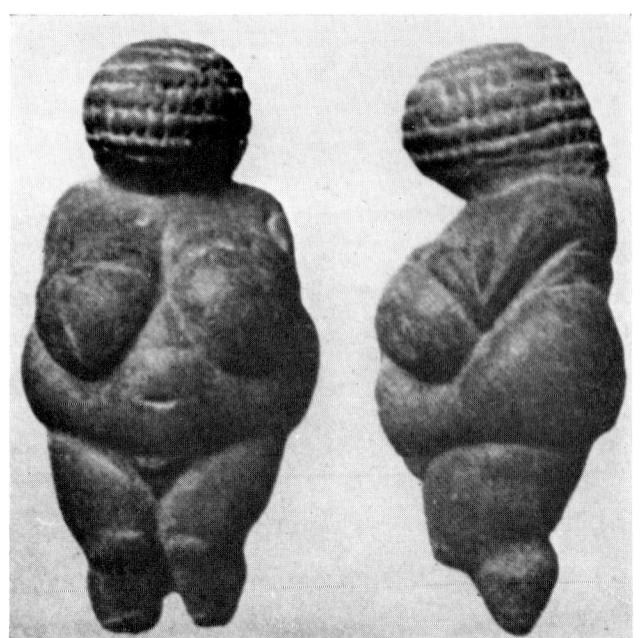

35. Venus of Willendorf. Adipose type. From Boule, *Les Hommes Fossiles*.

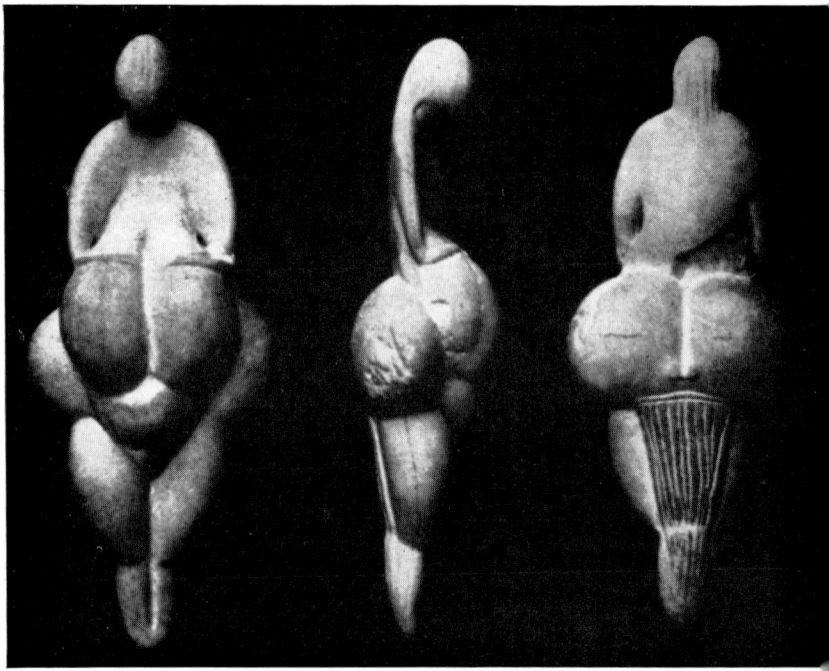

36. Venus of Lespugue, steatopygous type. From Boule, *Les Hommes Fossiles*.

# CHAPTER 16

# Tumours

Tumorous growths are difficult to incorporate in a system as their aetiology is still unknown. We shall discuss them here as they can be regarded both as a reactionary process, and a metabolic disturbance, or parasitosis. A growth does not show an egoistic purposeful self-development: it tends to partial self-destruction, as the possible crumbling away of a 'cauliflower excrescence' illustrates, and what is more it leads to quick destruction of the host organism, and consequently of itself. This is the essential difference between a tumorous process and parasitosis: although the latter may cause its host's and its own destruction the parasite has the chance to preserve the genus by transferring the infection, or else the host is destroyed and the parasite sporulates, as happens with tetanus or splenic fever, so that this latent life may flower again when conditions are favourable.

Evidence of tumours is rare and difficult to interpret. Benign tumours, osteomata, are sometimes difficult to differentiate from the final stage of an inflammatory process. I have already mentioned that Esper wrongly interpreted hypertrophic callus as osteosarcoma. The oldest known tumour is an osteoma of a vertebra, discovered in a Mosasaur from the Cretaceous, the Platecarpus, found at Niobrara in Kansas (124). Two other cases are recognizable on Neolithic femurs, and one of the ribs from the human remains of Furfooz has an osteoma (Pl. 38).

The terminology used by several authors causes difficulty. Osseous growths or neoplasms are not always given the right term. Well defined knob-shaped or flat perioseal bone formations of an inflammatory kind are called osteophytes. Exostoses are outwardly defined new bone formations which show more correspondence to real tumours (Moodie) (124). Herman and Morel have formulated the following definition: 'These tumours develop either in a physiological way at the expense of the periosteum (exostosis) or at the

expense of the marrow or the cartilage (central osteoma) or by hetero-
plasia or metaplasia of the connective tissue.'

The tumour on one of the above-mentioned Neolithic femora
which was described as an osteoma, is in fact an exostosis. Indeed
the radiograph shows clearly that the periosteum was the starting
point.

Moodie regards a tumour of the caudal vertebrae of an Apatosaurus from Wyoming as a haemangioma, because it was highly
vascularized. Nevertheless he is careful not to deny that the tumour
may have been the result of chronic osteomyelitis or that it could
even be callus around a fracture.

Some cases of dental tumours have been noted without, however,
certain diagnosis. It is worth mentioning the ivory-like tumour on
a fossil elephant's tusk from the caves of Paviland (South Wales),
with the base partly lodged in a cavity.

In connection with malignant tumours Abel records a case of
osteosarcoma in a bear from the cave of the Dragon at Mixnitz.
Details about the case are missing and he himself calls his diagnosis
a 'probable' one. It is the same with a similarly described affection
on a long-bone of a horse from the cave Pair-non-Pair. The bone was
found by Daleau and dates from the Pleistocene period.

Three cases of osteosarcoma have been noted from Egypt: one on
a femur and two on humeri of individuals who lived during the
Vth dynasty, they were found in a graveyard of the pyramids of
Gizeh. The same disease was diagnosed *per exclusionem* on a highly
vascularized tumour of the pelvis found in the catacombs of Kon-el
Shougata in Alexandria (*c.* A.D. 250).

Growth formations are also known from the Pre-Columbian
period. MacCurdy (113) says of a skull from Pancarcancha that it is
'as picturesque as it is gruesome'. It belonged to a man of sixty and
has an enormous osseous growth on the cranial roof, caused by
osteosarcoma (Pl. 39). The growth is rooted in the left parietal
bone, grips the frontal bone, and, spreading over the sutura sagit-
talis, affects the right parietal bone. The growth has an average
thickness of about 45 mm. Underneath it the cranial wall has completely disappeared. Even outside the limit of the osseous growth
the bone has been affected over a great area, so that the total surface
forms an oval of 14 cm. by 11 cm. The development of this growth
can have lasted many years.

Pales (135) describes a humerus from the dolmen of Meudon
on which the distal epiphysis and the corresponding part of the
diaphysis show irregular osseous outlines. The surface is sown with

holes, which are a sign of profuse vascularization. The volume of the bone is magnified and it seems inflated, in particular at the level of the substantia compacta. It is hardly possible to discern the medullary canal. These points and the fact that the lesion is situated in the epiphysis compel us to regard it as a myeloplastic growth.

Influenced by Cushing's studies of a meningeoma, Moodie has described a fairly common neoplastic growth on skulls of old Peruvians as being one of these tumours. The general shape of the growth, its spongy appearance and its position on the left side of the skull led him to this diagnosis. As this disease does not occur on the European continent, Pales does not hesitate to connect it with symmetrical osteoporosis. Indeed he regards both pathological conditions as morbid characteristics of one and the same kind. At the same time he sees a direct connection between artificial cranial deformation and irritation of the cerebral membrane, as the causal factor of meningeoma. I shall not enlarge on Pales's argument because American anatomo-pathological pieces are very difficult to date and these cases do not form a point of connection with our conceptions about European Palaeopathology.

Wells (188) describes a case of carcinoma on a skull from the IIIrd–Vth Dynasty of the Old Kingdom of Egypt. The disease has affected the rhino-pharynx, resulting in the destruction of the upper jaw, the palatine and pterygoid bones, and has spread over the whole skull in the form of circular holes, between 2·5 and 9·6 mm. diameter. The signs of osteitis around the primary lesion are only of secondary importance, and originated from ulceration after the malignant growth had broken through the mucosa of mouth and nose. He also points out that several of the few malignant tumours in mummies, found in ancient skeletons are indeed tumours of the nasopharynx. These are still a common occurrence in East Africa, in particular among young people (Pl. 40).

The rare occurrence of carcinoma in prehistory could mean that cancer is a disease of civilization. But this is not really true: the cases mentioned are of sarcoma, which mostly occurs in young people. Carcinoma, more strictly, is a disease of old age. I have already referred to the low average age of death, which is much lower than the predilection age for cancer. This disease only came into the foreground with the increased life-span attributable to civilization, so that the chances of a cancerous affection have increased. Another important point is that primary osseous neoplasmata are very rare, and consequently it is difficult to recover these anatomo-pathological pieces. This even applies to metastases:

in general mainly breast, womb or prostate carcinomata cause metastases in bones. In some cases condensation of the bone then ensues, but more often it is decalcification and necrosis which in turn cause spontaneous bone fractures. Metastasis sometimes results in the complete disappearance of bone. The difficulty here, when examining, is to find out whether the disappearance of the bone is due to a pathological process or to an artificial one, arising after death. Of course diagnosis can be based on the reactions in surrounding bone tissue, which can be inflammatory but also reparative. It is important to realize that fossilization and mineralization of bones can be detected through radiographical examination, but that they should be identified through microscopic examination, which in many cases involves great difficulties.

# CHAPTER 17

# Reactions

The most important characteristic of a living organism is what we call the instinct of self-preservation. This takes on a positive character when seen as a part of the preservation of the species, though it has its seemingly negative counterpart in the instinct of self-preservation. The latter shows itself not only in conscious reactions of the organism, such as flight reactions when in danger or aggressive and defensive reactions when attacked, but also in unconscious responses within the organism itself, which we refer to as neuro-vegetative conditions. They are important in the field of pathology and traumatology. I have discussed them briefly in connection with fractures, because of their importance. These unconscious reactions can be instant and violent, such as vomiting and diarrhoea after bad food has been eaten, or sneezing and coughing when the respiratory canals are stimulated. Other reactions are slower though they arise directly from the stimulus – for instance infection, fever, and recovery. Sometimes they are not even noticed by the patient – for instance immunity and anaphyllaxis.

IMMUNIZATION

Tracing the last type of reaction in Palaeolithic and Neolithic finds has been impossible up to now because the soft tissue is missing. I know of only one exception: the mammoth, found in Siberia, preserved in ice, unchanged for 175,000 years. Precipitation reactions of its blood gave the same results as those of an elephant today. The study of immunization may be possible for Egypt, although there is one difficulty: if we are dealing with embalmed mummies, very seldom shall we find enough dried blood to do the necessary laboratory tests, because they have been kept in brine for a long time. Williams has recommended testing blood reactions only on those mummies which have been subjected to a desiccation process. W. C. Boyd and L. G. Boyd succeeded in ascertaining blood groups from muscular and osseous tissue (161).

Moodie (124) describes fossilized red blood cells in bones of Apatosaurus. Blood cells had already been described in the remains of the Iguanodons of Bernissart. He explains this phenomenon as follows: the film of red blood cells contains cholesterol which turns into palmitate or stearate salt. These salts preserve the shape of the cell and crystallize.

INFECTION

We are interested primarily in reactions attacking the osseous tissue, this being the most important material. These reactions are caused by physical influences, bacterial toxins, and unknown stimuli such as arthritis and tumours. The latter have already been described in a previous chapter.

The result of these reactions is inflammation, in the broad sense of the word, which can be sterile in case of physical irritation. According to the locality in the bone we speak of periostitis, osteitis, and myelitis. Usually the zone is not strictly confined to the osseous membrane, to the osseous tissue, or the marrow, so that the term osteomyelitis is used more often. Even if we cannot trace the causal agent, still the existence of these reactions allows us to consider localizations other than in the bone itself. Indeed the condition of the bone usually results from a primitive focus in some soft-bodied organ, for instance tonsillitis, tooth caries, furunculosis, pneumonia or typhus, and certainly from tuberculosis and syphilis (at least if we assume that the latter occurred in Palaeolithic times).

Periostitis often develops after bruising of the periosteum. The aseptic reaction subsides quickly after a previous thickening of the periosteum. In chronic cases not only does this stimulus cause a thickening but also new formation of bone, so that the surface becomes uneven and rough. We find examples in Egypt, from the Neolithic period, in cave bears, and even on the humerus of a Mosasaur from the Cretaceous found in America.

Osteomyelitis has been found in the fractured vertebrae of a reptile in Texas, from the Permian, presumably Dimetrodon of the type Edaphosaurus (124). The condition also occurs in Pleistocene animals, such as the bison, cave bear, and several carnivores. A special variation of osteomyelitis is mastoiditis which occurred a lot in Egypt, Nubia, and Pre-Columbian America. The localization of the infection allows us to assume either an infection of the respiratory tracts or an infectious disease, such as measles, scarlet fever, or typhus. In Palaeolithic times no cases of osteomyelitis in humans are known, unless we accept that the puzzling skull from Broken Hill, with Neanderthal characteristics, belongs to the Palaeolithic period. This skull, the lower jaw of which is missing, has fifteen teeth of which ten show very bad caries, and usually the Palaeolithic period is taken to supply no absolute proof of the existence of caries. The roots of the teeth show traces of abscesses. Moreover, there are two perforations situated in front of the left ear as a result of the abscessed petrous bone. Yearsly thinks this must have caused death through pus sinking as far as the squama temporalis and neck into the chest-cavity (3). Other writers (100) assume that the lesion was caused by the teeth of a carnivore. In the Neolithic period, however, osteomyelitis does occur in our regions, and even more in America during the Pre-Columbian period. A lower jaw (35) from the burial cave of Furfooz (province Namur, Belgium) shows a typical lesion of a healed osteitis, which had originated from a tooth abscess (Pl. 41). The inflammatory process caused important change in the condyles of the mandible, so that the mastication mechanism must have certainly undergone important changes after healing. I discovered the same process in a rib of a member of the same population (93).

Remarkable is the Peruvian skull, tormented by pansinusitis (Pl. 65, chapter on trepanation). The forehead shows an enormous fistula. Underneath, the bone is rough due to reaction, caused by the chronic suppuration of years (124).

MacCurdy (113) describes a fracture of a left elbow with osteomyelitis. Humerus and ulna have fused together completely, but the radius has stayed free. A little above the elbow joint a rather large piece of bone is to be seen in an oblong cavity. It is the sequestrum. The widening of the humerus halfway along the diaphysis is a reaction to get free from this sequestrum.

In the Bible we read (Job ii, 7 and 8): 'So went Satan forth from the presence of the Lord, and smote Job with sore boils from the sole of his foot unto his crown. And he took him a potsherd to scrape himself withal; and he sat down among the ashes'; and further (Job. xxx, 30): 'My skin is black upon me, and my bones are burned with heat'. The description corresponds to one of osteomyelitis, originating from a skin infection, which itself seems to be a secondary development of eczema. The soothing quality of ash is remarkably like that of talcum powder, which we use nowadays.

Moodie (124) devotes a special chapter to the skeletons of dinosaurs, small pterodactyls and archaeopteryx, which have been
found in a position of opisthotonos. He regards them as the proof
of a neuro-toxic condition, rather than of normal death-throes,
which means that these animals died of an infection of the central
nervous system. In Moodie's opinion, the infection might have come
about easily, because the brain of this sort of animal was little protected by bone or teguments. This argument does not seem very
convincing to me.

# CHAPTER 18

# Arthritis

It still seems impossible to give a definition of arthritis as a patho-logical entity. We can hardly begin to give its aetiology. For this reason it seems better to devote a separate chapter to the subject, rather than to class it under some other chapter of this work. Sigerist, on the other hand, discusses this affection under the heading of infectious diseases because he, being an American, accepts the infectious aetiology of arthritis.

The connection between tooth caries and arthritis is not in fact an American discovery. This is proved by one of the Assyrian letters from the Kuyunjik collection (eighth century B.C.) which reads: '... the inflammation which grips the man's head and neck, and wrings the joints of his arms and legs, comes from his teeth. These teeth of his must be drawn. They are the root of his nagging pain...' (191).

We shall discuss arthritis in a more or less logical order, so that the aetiological connection with other pathological conditions will show up more clearly. There is of course one disadvantage, namely that such a system is too artificial. But it enables us to bring to-gether and compare a group of phenomena which sometimes are identical but get muddled because of the wide variety of the names used, and aetiological factors seem to be one of the reasons for this. It is not at all surprising that in Pales's work (135), for instance, we find a description and discussion of lesions which now seem too artificial. This is understandable. Pales gives an extensive 'explana-tion of principle', which, because of the results of research in recent years, can no longer satisfy us, even though the introduction of cor-ticosterone in the therapy has not made the aetiology of arthritic affections any clearer.

It seems simpler to start from one particular arthritic affection, to give its characteristics, and to compare it with other similar diseases. In this way we can make a distinction between several separate conditions of the disease, but at the same time we may find that the aetiological factors are so different that we can hardly treat the distinctive patterns of the disease within one homogeneous framework.

If we go back in history about 150 years, we see that several pathological conditions were separated from the overall picture of disease of polyarthritis, because they had come to be described and recognized as separate, clearly defined diseases (62). In 1836 Bouillaud described a disease which is called after him, so that the remaining group was reduced to chronic evolutive polyarthritis and arthrosis. In 1850 Garrod differentiated gout as a distinct disease and reduced the rest of the group to chronic evolutive polyarthritis and arthrosis. In 1898 Bannatyne managed to separate osteoarthritis (osis) from this group and only chronic evolutive polyarthritis remained over. Nevertheless this disease should not be regarded as a specific disease on its own. Indeed some syndromes can be accompanied by polyarthritis, although this cannot be regarded as 'real' polyarthritis, because symptoms other than those of the joints come to the fore, such as a high fever, localizations in the heart and lungs, in the lymph nodes, the skin, mucous membranes and spine, together with splenic symptoms. Furthermore the age at which this disease comes about, or the course it takes, may not correspond with those of classical chronic evolutive polyarthritis. The best known of these forms are psoriasis-arthritis and spondylitis rhizomelica. There also exist transitional forms between the above mentioned groups and the classical chronic evolutive polyarthritis. In the latter, analogous anomalies occur as a result of inflammation processes, not only in the joints, but also in the peripheral organs, such as muscles and aponeuroses. Nevertheless a further division based on anatomopathological or laboratory data for chronic evolutive polyarthritis is not possible, and the above mentioned syndromes are best regarded as specific forms of the disease, even though not typical. We can only speak with reservation about spondylitis rhizomelica, Reiter's disease, and certain syndromes which are accompanied by lesions of psoriasis.

Since, as has already been mentioned, the aetiology is still unknown, chronic evolutive polyarthritis is diagnosed on the basis of a certain number of clinical and laboratory tests, in which polyarticular symptoms occupy the central place – to this have to be added a number of localizations in connective tissue and other organs. If this combination occurs in certain affections, where the aetiology is known, as in some acute and chronic infectious diseases, chronic evolutive polyarthritis would only be a collective noun for certain affections of the connective tissue, a systemic disease of the mesen-chyma. But certain other syndromes occur, which can also be re-garded as systemic diseases of the connective tissue and of which the aetiology and pathogenesis are not at all clear. Klemperer classified these together with the group of the 'collagenoses'. The following are regarded as collagen diseases: Bouillaud's disease, chronic evolutive polyarthritis, generalized erythema lupus, peri-arteritis nodosa, sclerodermia and dermatomyositis. The first mentioned affection is becoming better known, and tends more and more to be given a separate place outside the group.

In an extensive study Pales (135) takes up the whole range of aetiological factors and, based upon his personal opinion, draws interesting conclusions. The environment seems to be very im-portant: 'chronic arthritis is in a certain sense the fruit of civiliza-tion'; and Baudouin does not hesitate to ascribe it to 'a microbe, unknown up to now'. He was writing in 1930. 'This microbe was transferred from man to the animals, through domestication.' This certainly is in contradiction to the fact that arthritic lesions are already to be found on very primitive reptiles in the Secondary geological era.

Cold and dampness have for centuries been regarded as a con-tributory factor involved in the onset of arthritis. Mozes had studied the degree of dampness in his house, and those of his students. De Mortillet regards it as having caused arthritis in the cave bear.

I cannot agree completely with these opinions. In the first place these lesions have been found on a crocodile in Egypt, dating from the Miocene period, when the climate was warm, and comparable to climates we now know in Malaga, Madeira, Southern Sicily and Southern Japan; and also Ancient Egyptians suffered from these lesions, which they contracted, according to Wood Jones, in the water of the Nile. But furthermore, mammals in America from the Eocene period, with its soft climate, were affected and, most important, reptiles from the Secondary also show these lesions – 'reptiles, which are great friends of warmth', as Boule says.

Michez (122) has taken up again the importance of climatic in-fluences on the origin and evolution of disease. He comes to the conclusion that disease occurs more frequently in cold and damp regions with great fluctuations in temperature and humidity, than in warm and dry regions. This influence does not necessarily have a direct result. Its effects may be contributory.

Thus we can assume that a difference exists in microbial flora and in their virulence. On the other hand we could suppose that the spraying of fine waterdrops promotes the spreading of microbes.

Yet we do see that microbe infections are of a more serious kind and also occur more frequently in hot regions than in colder ones, without the natives ever acquiring non-specific polyarthritis, provided they stay in their country. They do however become susceptible when they move to our regions and take part in hard manual labour.

This raises the question of whether there exists a racial or hereditary disposition to the development of chronic evolutive polyarthritis. Stecher (162) answers this positively and says that this disease is inherited as an irregular autonomous dominant factor. It occurs six times as often in the familial community as in a whole population.

Michez explains climatic influence, and particularly sudden climatic fluctuations, in an indirect way. Man has to rely too often on his thermo-regulating ability: quick successive vagal and lymphatic stimuli may exhaust the adrenergic and acetylcholinic reactions. This may give rise to abnormal metabolites which may cause an inflammation process in a non-protected cell.

Another climatic factor is sunlight. Its action may possess anti-inflammation qualities when taken in moderate doses, but over-exposure to it causes inflammation. The healing influence of the sun is explained by the presence of ultra-violet rays, which, according to some scientists, stimulate the corticotropic action of the pituitary gland, and according to others transform the skin steroids into substances with a corticoid effect.

Finally Michez thinks that vagotomy plays a part in the origin of chronic evolutive polyarthritis.

The age factor has not been neglected. Pales maintains that, in general, people who are affected have reached a certain age, and so also in prehistory. This seems contradictory to the facts we have about the age of these people, and to the fact that arthritis also affects young people nowadays. Vallois (170) on the other hand claims that arthritis occurs at an earlier age in prehistoric man than in man today, that is, before the age of thirty.

Most attention has been given to trauma, and, indeed, still is. Because of the caudal localization in the dinosaur, Osborn thought this was due to the important part the tail plays as a support. Moodie (124) rejects this view because other segments are affected too. He prefers the idea of the possibility of a sprained ligamentum vertebrale commune from using the tail when swimming.

Pales has studied the predilection places of arthritis in several animals and humans, fossilized ones as well as recent civilized and uncivilized ones, and reaches the conclusion that first those segments in which the individual does most of his movements are affected. That is why prehistoric man is attacked by the disease mainly in the front part of the lumbar area, the active segment in a physically active individual, and modern man is attacked in the cervical part of the spine, because of his sedentary existence. Pales makes a distinction between men and women, so that cervical spondylosis is characteristic of the fine neck of a woman, and lumbar spondylosis of a man, as a result of his occupation. He also devotes his attention to lumbar arthrosis, a disease described by Léri and characteristic of soldiers living in trenches where the damp climate together with the traumatic factor of carrying a rucksack have a harmful influence.

Pales even takes the argument so far as to find a connection between the position of the lesions on the vertebrae and the dominating movements in each segment. In conclusion he repeats the thesis that rarefaction of the bone is compensated by ossification in the ligaments. He bases his argument on the symptoms of senility in 'individuals who lose their phosphates'. In this he includes tabetic spondyloses and diabolo vertebrae corresponding to those of chronic arthritis, as well as pregnancy, beside which there is an important decalcification factor and a changed static factor, so that the lumbar lordosis becomes more pronounced. The influence of the cold, he says, lies in the fact that through vasoconstriction of the skin the spine acts as a sponge and hypervascularization of the spine then promotes ossification. He considers that the influence of the nervous system, which expresses itself in pachymeningitis, marrow lesions, and bruises, has a trophic influence, which should not be underrated. We could regard this explanation as parallel to Michez's theory, though the mechanism is different.

Finally Pales repeats that not only has spondylosis preferred localizations, but also that these are the same for infections, with the difference that tuberculosis affects preferably the lumbar segment, and syphilis the cervical one.

One of the pieces studied consists of the vertebrae of a cave bear with ankylosis involving an important kyphosis, which Abel considers to be of traumatic origin, in the form of a fracture of two vertebrae during youth. Pales describes a similar lesion, and Moodie also mentions the lesion in a Pliocene horse, and a Pleistocene camel, and wolf.

Spondylitis ankylosans (5) is regarded by Bechterew as a stiffening of the spine, progressing downwards from the top. It always involves kyphosis of the dorsal side of the spine. Big joints are not affected and radicular symptoms are always present. Strümpell and Marie have described cases in which the disease first affected the caudal part and progressed upward, and in which spinal malformation did not occur, although the big joints, in particular the proximal ones – such as the shoulder – and hip-joints, were affected (spondylorhizomelia). In chronic vertebral arthritis malformation of the corpus vertebra occurs, while in Pierre-Marie's disease they stay cylindrical. They are abnormally transparent, especially on the outmost parts. This difference led to the name of 'Bechterew's disease' for spondylitis ankylosans, and when the big joints were affected to 'Bechterew's disease type Pierre-Marie-Strümpell'. This division seems to be very well grounded nowadays.

Bechterew's anatomo-pathological research work convinced him that an inflammatory process of the cervical membrane was involved, causing paresis and atrophy of the muscles of thorax, back, and shoulder. These factors have primary importance. The degeneration of the vertebrae is a secondary factor. Muscular atrophy is at the base of later kyphosis forming. This causes compression, followed by melting away of the intervertebral disc, which leads to ankylosis of the vertebrae. Ehrhardt has noticed ossification of the hip-joint ligaments, as well as of the paravertebral. Léri and Marie regarded spondylitis ankylosans as an infectious or toxi-infectious osteopathy with inclination to osteoporosis, where ossification is a recovery symptom produced to support the porous and breakable osseous tissue. Sivén and Fraenkel believed that it was an affection of the joints of the spine only. Güntz also thought it was in the first place an affection of the small spinal joints, that is to say inflammation with suppuration and infiltration of round cells, hyperhaemia and morbid formation of connected tissue. This is the origin of the name spondylarthritis ankylosans. These lesions of the small joints form the difference between spondylitis ankylosans and spondylitis deformans. The latter was regarded primarily as a degeneration anyway. In 1930 however Ehrlich came to the conclusion that the two affections should not be separated. Simmonds regarded calcification of the apparatus of the spinal ligaments as most typical of Bechterew's disease. Klinge (5), (1934) saw the disease as a form of arthritis. Other authors hold the same view and it still seems sound today, though Van Swaay (5), (1950) does not agree that the disease is of an infectious nature, and considers a lesion in the cartilage as the most important cause.

Since English physicians regard it as an inflammatory process, they call it ankylosing spondylitis. American physicians call it rheumatoid spondylitis because they see a close relationship with arthritis. Aufermauer (5) ascertained (1949) that spondylitis ankylosans is characterized by ossifying inflammation of intervertebral discs, ligaments, joint capsules of the spine, and by ossification on the surface of the joints.

Pathogenically we can say that clinical evidence points mainly in favour of the theory that spondylitis ankylosans should be classified in the group of chronic arthritic diseases: the jerky evolution, the passing pain in the joints, increased sedimentation rate, fever, sometimes present, and raised leukocytosis.

Anatomical factors point the same way: corresponding inflammatory degeneration symptoms in the intervertebral joints, with hyperthermia accumulation of round cells, and morbid growth of connective tissue. Since ossification is the most conspicuous symptom we can assume spondylitis ankylosans is a chronic ossifying arthritic infection. There are however remarkable differences between spondylitis ankylosans and chronic joint arthritis, as has been made clear by Pugh:

|  | CHRONIC JOINT ARTHRITIS | SPONDYLITIS ANKYLOSANS |
| --- | --- | --- |
| *Predilection of sex* | occurs more in women than in men | more in men than in women |
| *First occurrence of disease* | between twenty-five and forty years | between fifteen and thirty years |
| *Calcification of ligaments* | seldom | general rule |
| *Radiograph* | rarefaction of the bone under cartilage | sclerosis underneath cartilage |
| *Skin nodosities* | often | never |
| *Iritis* | seldom | often |
| *Serology* <br> a) agglutinin for haemolytic streptococci | often | seldom |
| b) plasmaphosphatasis | normal | increased |
| *Improvement of disease* | seldom | often |
| *Chrysotherapy* | often effective | not effective |
| *Radiotherapy* | without result | effective |

To account for this sickness any number of causes have been appealed to: age, infection, tuberculosis, gonorrhoea, syphilis, cold and dampness, or trauma. The disease can be traced back to the Secondary period, where it occurs in very primitive reptiles. When the latter died out the disease survived. The alternating warm and cold periods have no influence on the disease, which is spread over the whole world. It has been assumed that the disease has proliferated, although this is impossible to say: the available material is too scarce to allow us to draw such a conclusion. The disease still occurs a lot in wild animals.

According to Aufermauer only two factors count: a traumatic one and an endocrinal one. The first factor does not explain the jerky course of the disease or its infectious character. As far as the second factor is concerned, he lays great emphasis on the parathyroids. The removal or ligature of these glands should improve the condition and even cause complete recovery. Others deny these facts. Post-mortem examination of the parathyroids did not, in these cases, show any deviations worth mentioning.

Pales (135) devotes his attention to spondylitis rhizomelica, a disease which, as has already been mentioned, starts in the lumbar region. The ankylosis causes pains, which disappear during the stiffening process of the dorsal segment and appear again when the process starts affecting the cervical part. The disease consists in calcification of the ligaments. The radiograph is typical: one can distinguish it from vertebral arthritis, because with the latter the corpora vertebrarum are deformed, while with Pierre Marie's disease they stay cylindrical, although they are abnormally transparent especially in the outmost parts. This is very pronounced on the spinous process. These symptoms have been found on Neolithic vertebrae. It is striking how often massive ankyloses occur in Peru. Moodie also found the disease on dinosaurs from the Secondary and Pales thinks it might be the above-mentioned disease, but does not want to confirm it with certainty because he has no further facts.

Spondylitis ankylosans – other than of traumatic origin or as Pierre Marie's disease described by Pales as a separate pathological entity – and ordinary arthritic spondylitis (Pales makes no distinction) occur from the Secondary onwards in the dinosaur (*Diplodocus marsh*) found in Wyoming. Other reptiles from the same period have similar lesions. I draw attention to the close connection between the frequent arthritis in the big dinosaurs, and the age of these creatures. They could reach an age of over 1,000 years, so that we can regard their arthritic affections as a real wearing out process. In

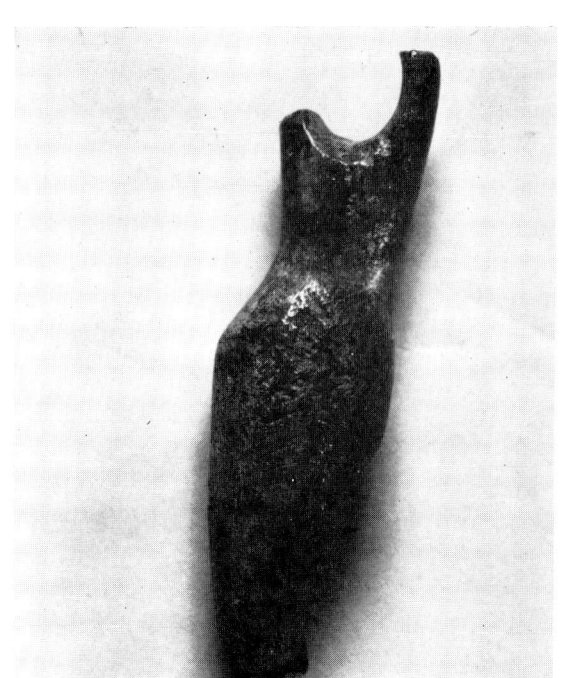

37. Venus of El Pendo
(Santander, Spain).
Prehistoric Museum,
Santander.

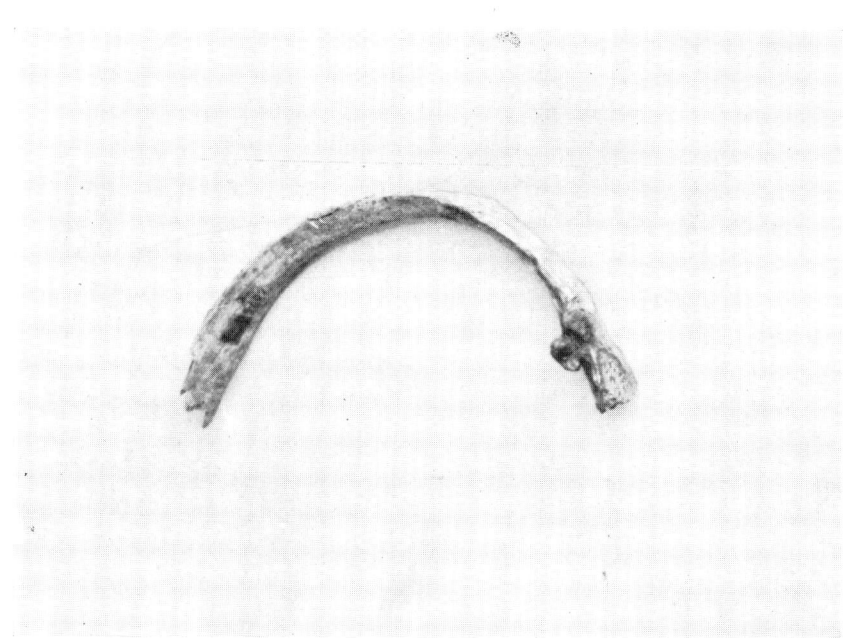

38. Osteoma of the rib. Race of Furfooz.

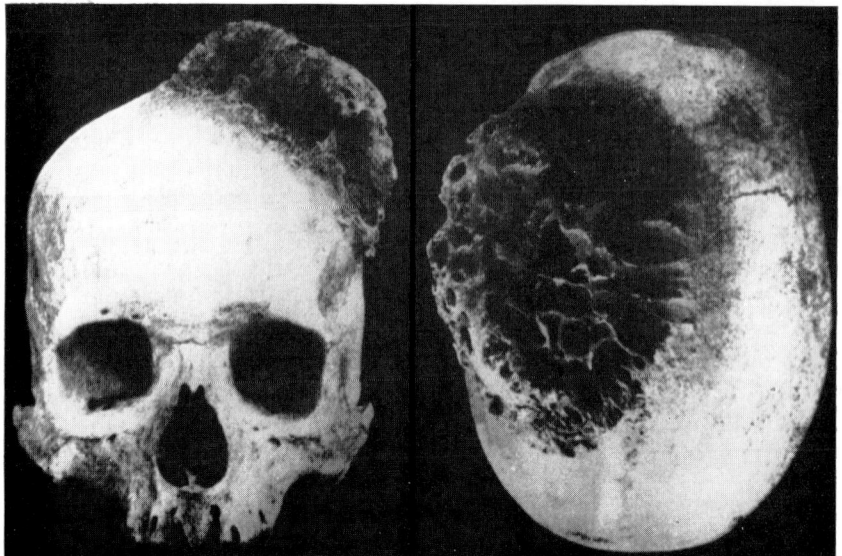

39. Osteosarcoma, described by MacCurdy. From Pales, *Paléopathologie et pathologie comparative*.

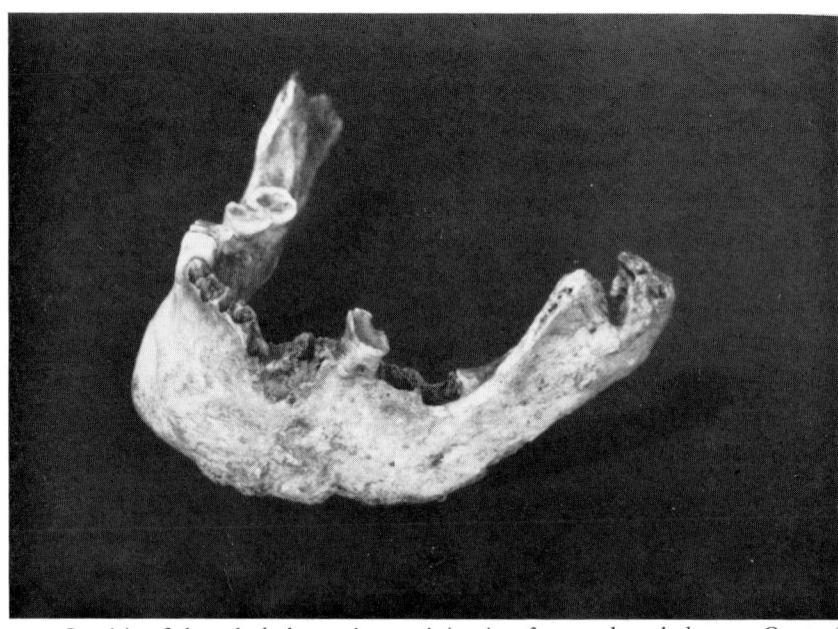

40. Osteitis of the whole lower jaw, originating from a dental abscess. Cave of Furfooz (Neolithic period).

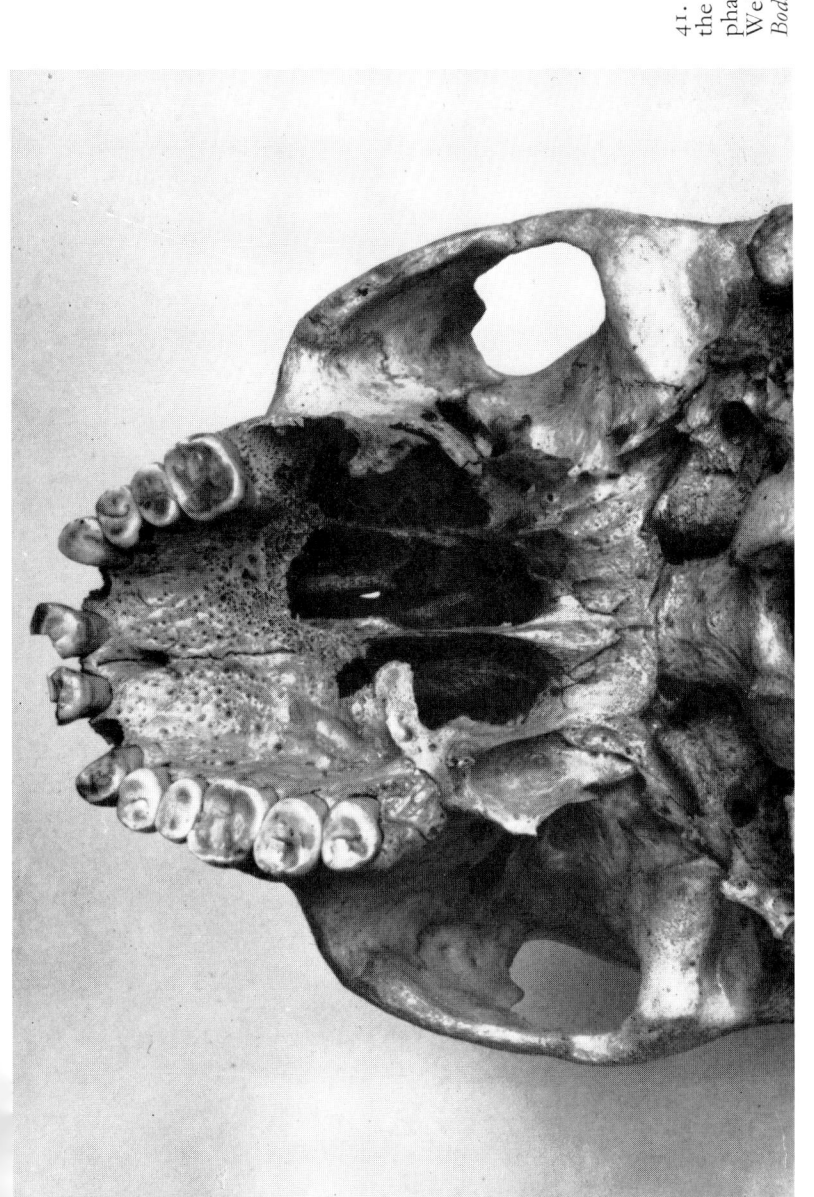

41. Carcinoma of the rhino-pharynx. From Wells, *Bones, Bodies and Disease*.

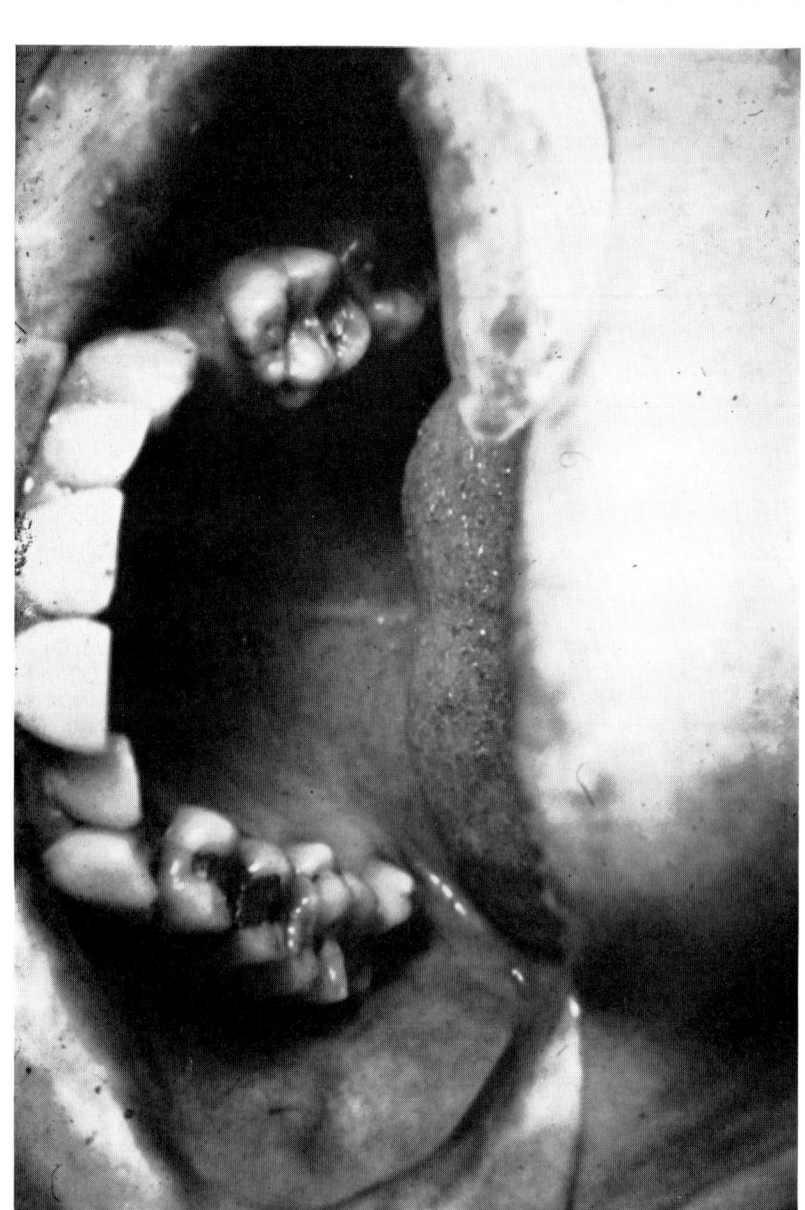

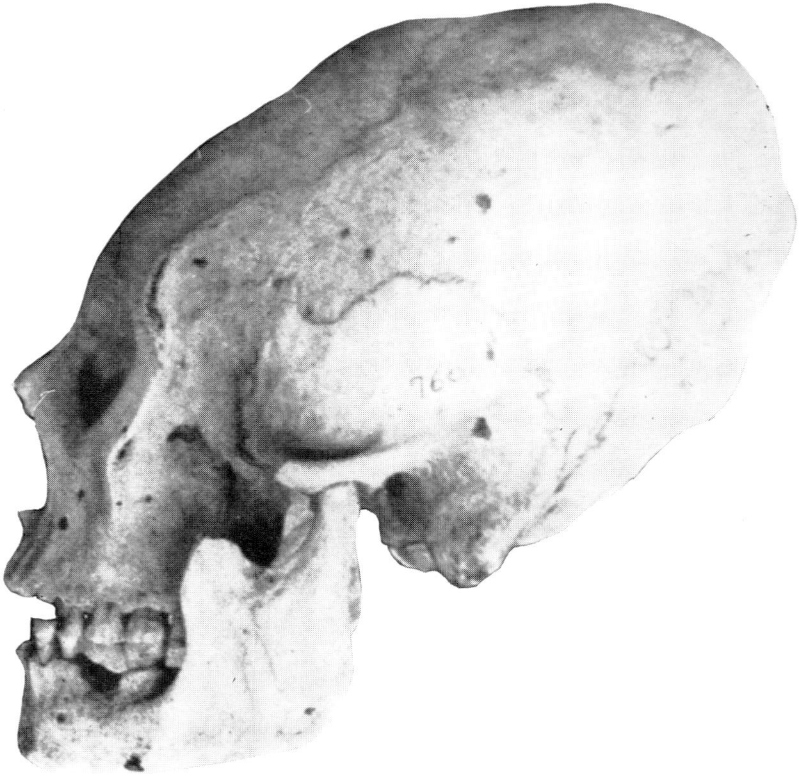

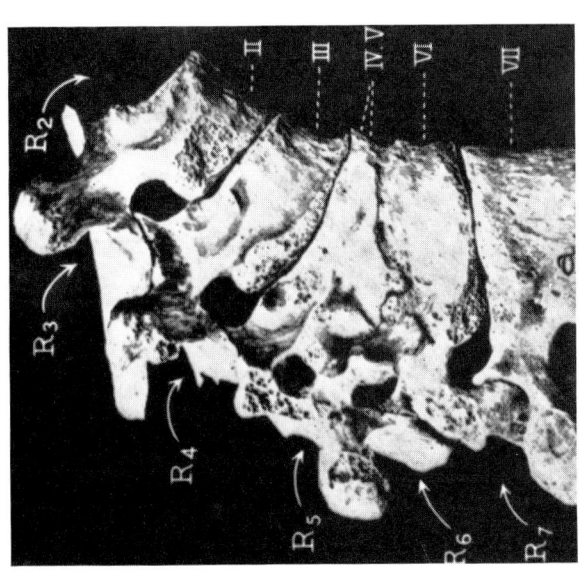

44. Pott's disease. Neolithic period. Heidelberg. Described by Bartels. From Moodie, *Paleopathology*. The roman numbers indicate the vertebral bodies.

43. Adult man with Aymara cranial deformation. Also alveolar abscess of the mandible. From MacCurdy (113). At Tronotoy.

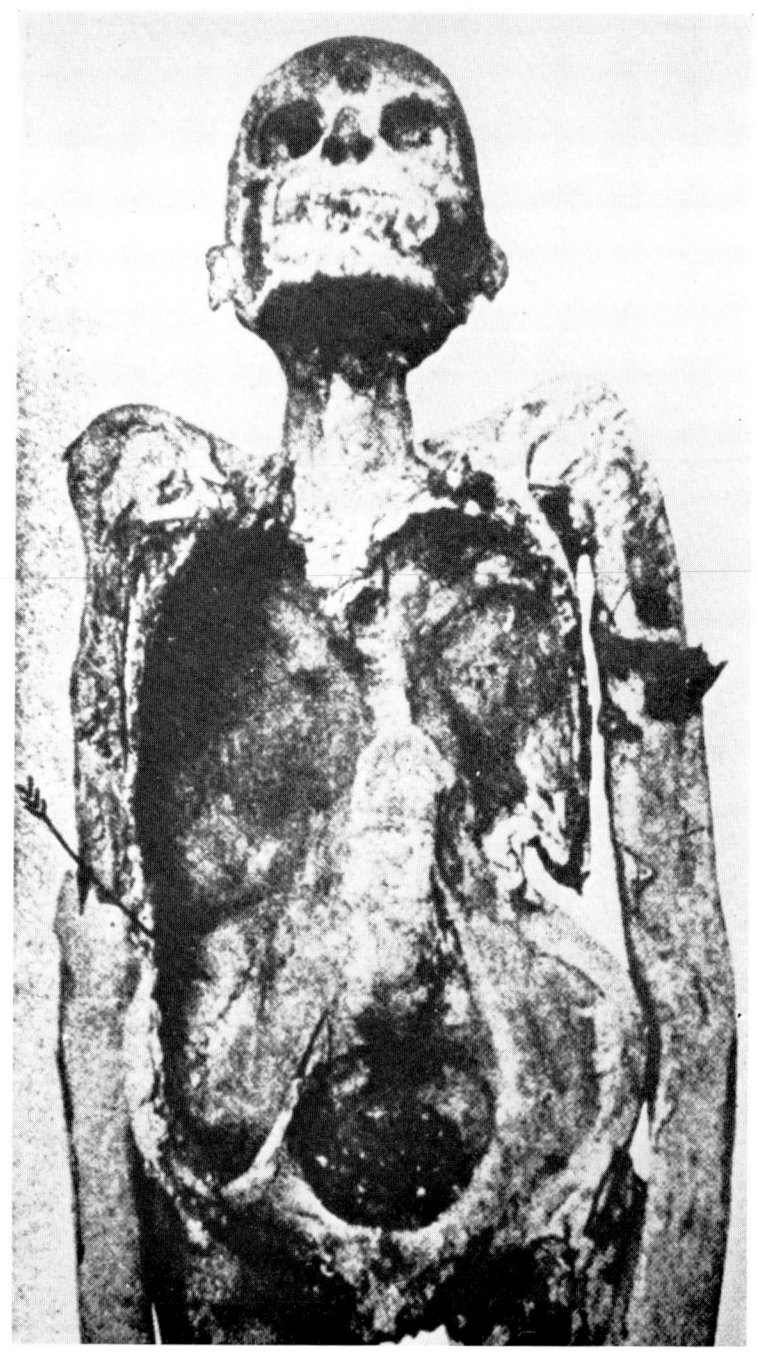

45. Pott's disease in a priest of Ammon. The arrow indicates the psoas-abscess. XXIst Dynasty, 1100 B.C. From Sigerist, *A History of Medicine*, vol. I.

the Tertiary the disease already occurs in primitive mammals. To this period also belongs the above-mentioned crocodile from the desert lake of Marinet. The number of known pathological cases from the Quaternary is relatively higher, as more fossils are found from this period. Sauria, canidians, felidians, ursidians all show the disease.

The Neanderthal man (of La Chapelle-aux-Saints) shows considerable lesions: the corpora vertebrarum are flattened, in diabolo, and covered with osteophyes. They are founrd particularly in the lumbar area, but also in the cervical and first dorsal vertebrae. Although these dorsal vertebrae must have contributed to the typical posture of this creature, they were not of primary importance for it. Its specific posture is mainly due to the placement of the foramen occipitale in the skull as well as to the structure of the spine. Spondylitis alone cannot account for this posture, but it may have aggravated it. To this should be added the effect of coxo-femoral arthropathy on both sides.

An Aurignacian skeleton from Solutré also shows this affection. A spinal column from the *abri-sous-roche* of Cro-Magnon is affected by osteophytic spondylitis over its whole length.

A lot of cases (144) are known from the Neolithic period and also from eneolithic graves (53). Osteophytes on vertebrae are even recognizable in cremation remains, as I was able to ascertain during the examination of the bone remains from the urnfield burial place of Neerpelt (80).

The disease occurs in Egypt in all periods. According to Ruffer 40 per cent of the mummies of the predynastic period were affected by the disease. Also the Pre-Columbians suffered from it. Hrdlička finds that 5 per cent of the Indians of Louisiana and Arkansas have arthritic bone lesions.

A clear case of extra vertebral localized lesions is the classical case of morbus coxae senilis of the Neanderthal man of La Chapelle-aux-Saints. This affection in classical form seems to have occurred, frequently in prehistoric times, as is shown by the examples for the Neolithic period, Ancient Egypt, and Pre-Columbian America. This frequent occurrence, and in particular the many bilateral cases, raises the question whether they are not cases of congenital subluxation. Some anatomical facts point in that direction: almost always the collum femoris shows a lessened oblique position, and the top quarter of the femoral head is stretched to the top and middle part of the collum femoris. In some places this occurs so frequently that heredity is accepted. The lesions correspond completely to those of

83

G

recent cases after radiographical examination: the joint cavity shaped like a half lemon, the hypertrophied joint head which protrudes outward, and the collum pressed together in anteversion. According to Pales these lesions are mainly due to overburdening of the joint. He regards them as a sort of physiological arthrosis, which in broad outline corresponds to mandibulo-temporal osteoarthritis. To this we can add that osteoarthritic localizations in the extremities occur very rarely in prehistoric times (123).

Only one case of mandibulo-temporal osteoarthritis is known, namely on one of the skulls from Krapina. The disease occurs a lot in the Pre-Columbians. Today it is very rare. Using these two facts as a point of departure Pales has tried to find the reason for this geographical and racial distribution and questioned whether this affection has not been overlooked by several investigators. To trace the aetiological factor he studied the occurrence of the disease in the races today. He came to the conclusion that the disease does not occur in Peruvians today or in Egyptians. In Negroes it occurs in 4·5 per cent, while it occurs frequently in Melanesians. In the New Hebrides 5·6 per cent of the people are affected, of the New Caledonians 24·02 per cent and of the inhabitants of the Loyalty Islands 26·4 per cent. It is striking that the disease occurs almost exclusively as a singular affection, which means that while other localizations do exist they are much less frequent than mandibulo-temporal arthritis. The characteristic lesions consist of flattening of the joint face of the condyle with a corresponding flattening of the joint cavity. The whole is studded with small osteophytes. It is also characteristic that the teeth stay healthy. Attempts have been made to retrace the aetiology of the disease to an infectious process. Yet most infections first occurred with civilization, and old skulls also show the disease. The influence of environment does not seem to be important, for the affection occurs among natives living in mountains, as well as in those from the coast. The possibility of an anatomical factor has been studied: the joint cavity of primitives was assumed to be much shallower than that of modern Europeans, with the result that movements could not be as extensive. But, if this is accepted, we must face the question why Neanderthal man does not show any lesions of this kind, although the same anatomical fact applies to him. In rejecting this view, we should think of the function of the joint, for in this lies the probable answer: the whole day long the Melanesian chews fibrous plants such as sugar cane. His mandibulo-temporal arthritis may be the expression of a wearing-out process.

The existence of a typical disease, such as coxo-femoral or mandi-bulo-temporal arthritis, in fossilized pre- and protohistoric races does not exclude the existence of other lesions. A Mosasaur from the Cretaceous, found in Texas, had lesions of phalanges and meta-tarsus. Similar deviations have been described on a Pliocene camel and horse from Nebraska. In the Quaternary they occur in herbivores and particularly in carnivores. In the case of the cave bears, the dis-ease became fatal and caused this species to die out. No prehistoric human race was safe from the affection: the material is very extensive and every day new finds bring to light new cases of the disease.

Although the disease occurs in the most remote times in all human races, extinct or not, there exists one varying factor: localiza-tion. This is a function of physical factors, depending on the way of life of each race. We could almost say it depends on each individual life. Osteoarthritis is on one hand the expression of a wearing out process, and at the same time it supports this function (Pales).

In connection with this myositis ossificans may be mentioned. This disease seems to be rare in prehistoric men, as well as in men today. But on the contrary it occurs a lot in animals. According to Bland Sutton exostoses of the femur of *Pithecanthropus* seems to be such a case.

In some primitive reptiles fine bone formations have been found along the edge of the spinous process of the vertebrae which have been recognized as ossified sinews. These formations usually stretch from the middle part of the dorsal area to the lumbar area, the sacrum, and the front part of the tail. This ossification seems to exist in two-footed dinosaurs, as well as in quadrupeds of the type Ceratopsia. Yet, if all Sauropodes seem affected, the armed Dinosaurus was evidently free from the disease. Without quite being pathological, myositis ossificans of two-footed dinosaurs seems to be a normal evolution of the sinews toward ossification. It may be a sign of senility, as in the case of the Javan musk-deer.

Moodie (101) describes similar symptoms in a sabre-tooth tiger (*Smilodon californicus*): the lesion consists of a great mass of bone along the spine. It is impossible to take it for spondylitis not only because of its shape, but also and particularly because of its site of origin. The osseous mass is completely shaped like the muscle and actually takes its place. Sometimes it completely fills its canal along the spinal column. It does not occur along the cervical part or the dorsal part, but does in the lumbar area, where it develops at the expense of the sacro-lumbar muscle and perhaps of the long back muscles. The disease shows different degrees, from the affection of the small musculi intertransversarii to the whole muscular mass. The joints stay free and if there is ankylosis, it occurs on the ligamentum commune ventrale. Sometimes there is a hiatus between two bone masses.

Abel describes such lesions in a cave bear, though their position is different, that is on the ulna, on which stratified bone layers can be seen. Also in canidians similar lesions have been found on the radius: here the symptoms have been explained as the result of a local irritation, the rubbing together of the front paws.

It remains to ask whether it is a lesion of the bone at the point of muscular attachment which causes ossification of the muscle, or simply a sudden ossifying of the muscle. Fay and Le Count think the new formations do not develop at the expense of muscular tissue, which would make the name incorrect. The development takes place at the expense of the surrounding bone, probably because of some traumatic stimulus, and the process slowly incorporates the muscle.

In the first place the aetiology has been traced to trauma: insufficient protection of the lumbar area causes the disease to be located there instead of in the dorsal part, which is protected by the ribs (Rozenstirn, Moodie). Nevertheless most lesions occur in spots which have much muscle. From this it would be more sound to conclude that myositis ossificans is due to excessive muscular labour, which exercises a continuous pulling on the sinews, and they in their turn exercise a local irritation at the point of attachment to the bone.

Recapitulating we can say that arthritis occurs at all times in all animal species, in the cold glacial age as well as in warm Egypt. The occurrence in Egypt seems to be due to working in the Nile water. The disease affects all races without exception. Sigerist (161) points to the importance of focal infection, in particular pyorrhoea, which occurred frequently in Egyptians. The disease has been noted not only in the Cro-Magnon race but also in Neanderthals, who lived in a warmer interglacial period. The frequent occurrence in cave-bears caused Virchow to name it 'Höhlengicht'.

'Höhlengicht', together with many well-healed fractures, occurring in the cave-bear are regarded as signs of its domestication. Dr Baudouin supports his view with the first, Dr Joseph points to the second. Pales rightly rejects these, pointing out that osteoarthritis already occurred in the Primary, together with well-healed fractures. The domesticated animal is, medically, the worst cared for: if its flesh is edible it will be slaughtered; if it is not, healing is left to nature. Dr Hanck thinks the enormous muscular masses of the thigh are some sort of fixation system of fractures. This too is rejected by Pales: these muscular masses in the vicinity of a fracture are, because of their situation between bone fragments, often the cause of pseudo-arthrosis forming in particular when there is no traction or immobilization. Quadrupeds achieve immobilization by walking on three legs.

It seems, in any case, dangerous to talk of domestication in a period before the Neolithic. Domestication supposes some sort of symbiosis between man and animal, which degenerates into parasitism when the animal is killed for food. The assumed domestication of the cave-bear does not imply anything else but the direct use of this animal to supply the Neanderthal's wants of the moment and in a very simple way. The enormous cave-bear would not have been a frightful monster – not at all frightening to prehistoric man. They would even have shared caves, in which the bear spent his winter-sleep. The enormous reserves of flesh, fat, and skin on this animal made it very convenient at the right moment. It could easily be killed while hibernating; the hand axe seems to have been the likely weapon. The cave-bear was first of all herbivore: lesions of actinomycosis seem to point clearly to this. When the climate became colder in the last glacial period its winter-sleep became longer and longer and its teeth grew alarmingly, which hastened its extinction. Konrad Lorenz says that to him it looks as if two creatures have, so to say, domesticated themselves: cave-bear and man. In the cave of the Dragon in Mixnitz we find in the cave-bear all the transitions in this process that we know of in dogs when they were domesticated (192). Abel holds similar views.

We can take things even further: the moment when prehistoric man for the first time conquered an animal ten times as heavy and strong as himself, was not only a technical revolution (because of the use of a hand axe) but also a spiritual one. At that moment for the first time the idea of a cult, of magic, awoke in him: man felt as if he had been helped by supernatural forces in fulfilling a deed which, at first sight, he could never have fulfilled by himself. From this, according to some writers, originated the cult of the cave-bear. In the mountain of the Dragon, above the village of Vättis, rectangular stone chests have been found in which bear skulls were discovered with the long bones of legs sticking out from the mouth and eye cavities. From this it was assumed by some archaeologists that the Neanderthals treated the severed heads in a special manner. In the cave of Mas d'Azil a passage has been found with a similar culture, dating from the Mousterian: at the end of the passage lie two bear skulls on flat stones, one belonging to a male the other to a female.

Around these, arranged in the form of a horseshoe, lie other bear skulls, pelvic remains and remains of mammoths. In other caves the skulls seem to have been placed on poles, around which ritual dances may have been performed. Nowadays a similar cult exists among the Ainu of septentrional Japan, the Schilkas of the banks of the Yenissey and the Arctic and Asiatic regions. For these peoples the bear is a supernatural being: some ethnologists regard this as the reflection of certain customs of the Neanderthals. Sacrifice of the bear probably was the first religious act of humanity. A bear was worth more than a man and ritual graves for bears precede those for humans. At the same time sacrifice of bears is quickly followed by human sacrifice. In 1939 in the cave of Guattari, near San Felice on Cape Circeo, a human skeleton was found which had undergone exactly the same treatment as a ritually buried bear skeleton.

The bear occupied such an important place that later in the Pleistocene period, when the cave bear had already disappeared and a more sophisticated technique had made it possible to hunt other animals, it still played a big part in the magic of the Cro-Magnons. An example of this is the clay model of the skull of a bear in the cave of Tuc d'Audoubert.

Nougier, however, does not accept the existence of a cult (133). According to him the vast quantity of bones of bears found in some caves is explained by the fact that bears found shelter in these caves when death approached. They had, as it were, to force their way through the masses of bones, which left the impression that the skulls had been arranged in a double row by humans.

If we find that the anatomo-pathological lesions are the same for prehistoric times as they are now, we can infer logically that the pathogenic factor is also the same. An absolute proof of the identity of bacteria does not exist. Have pathogenic germs always been pathogenic? Indeed they are one of the earliest forms of life, and we cannot even make out whether they are vegetable or animal! Bacteria have been found in the Algonkian rocks of the Gallatin formation in Montana. Walcott has described a micrococcus from there which shows a marked degree of similarity to micrococci today. Renault has studied bacteria found in coal, from the Devonian to the Jurassic, in faeces from fishes and reptiles, in the stomach contents and teeth of fossil vertebrates. Differentiation is impossible because this proceeds largely from culture media tests. Only from the end of the Palaeozoic do we know of bacterial infections. Because of the existence of parasitism in the Devonian, Moodie (124) does not hesitate to believe that the phenomenon of phagocytosis came into being at an early date. Ruffer managed to detect Gram positive and Gram negative bacteria in mummies. It seems obvious that Pre-Cambrian micro-organisms should not be regarded as pathogenic. They probably played a part in rotting processes and they acquired their pathogenic function only when vertebrates appeared, in particular the warm-blooded ones, whose constant bodily temperature was probably the most important promoting factor. Bacteria found in mummies should not be regarded too readily as pathogenic germs, as the possibility of secondary inoculation during the several preparations should not be underrated.

While there may be doubt about the pathogenic origin of non-specific inflammatory reactions this is not so for specific ones. Here the germ always causes the same lesions, which can only be caused by that one germ. Typical examples are tuberculosis and syphilis. This does not alter the fact that tracing them in fossilized bones still presents great difficulties.

GOUT

I place gout and uricaemia among arthritic affections and not among metabolic disturbances for two reasons. First because some authors see a connection between the two affections, and second because the diseases are difficult to differentiate when one has only bones at one's disposal. This is perhaps the reason why so many authors do not mention it in palaeopathological studies. Yet the disease was known among the ancient Greeks and it was treated with colchicum. C. Wells describes the disease in a 'mummy of an old man with long hair and flowing beard' from an early Christian period (v. Wells, *Bones, Bodies and Disease*, p. 109).

Radiographically it is not easy to distinguish gout from osteo-arthritis or chronic evolutive polyarthritis, because both these diseases show a proliferation picture, spur formations on the tendon attachments and bone cavities. Bone rarefaction in the vicinity of joints and lessening of the space between the joints, progressing to complete ankylosis, are found both in inflammatory processes and in atrophy of the joint through disuse (104). Nevertheless the punched-out area is very characteristic. The notches are not surrounded by the pronounced sclerotic edge which is found with osteoarthrosis or in chronic evolutive polyarthritis.

Confusion may also arise with hallux valgus, multiple myeloma and cystic tubercular osteitis, which often accompany Besnier-Boeck's disease.

# CHAPTER 19

# Dental diseases

Bones are important documents in palaeopathology, because they can be preserved for an almost indefinite period. This applies even more to teeth: their hardness allows their preservation in conditions when even bones succumb. Although it seems logical that the study of prehistoric dental disease and deviations should be part of palaeopathology, we have to take into account the fact that dental surgery has a separate place in medicine. Far-reaching specialization has separated dental surgery from general medicine in such a way that we should call the study of prehistoric dental disease palaeodontology and leave it to stomatologists. For this reason I shall limit this chapter to generalities and to points of contact between general pathology and stomatology. For a more specialized study the reader is referred to M. Brial's article (21).

Pyorrhoea alveolaris has always occurred, being found in fossil animals, though animals are less subject than man to this infection. We find the disease in a Mosasaurus from the Cretaceous found in Belgium. The animal showed grave lesions of the left ramus mandibularis. A forerunner of the horse, Meryhippus campestris, living in the Miocene (more than 1,500,000 years ago) suffered from alveolar osteitis, probably originating from actinomycosis. This caused the roots of the teeth to become bared. Moodie (124) however has not found any dental deviations in American fossilized reptiles.

Man has suffered from this disease at least since the time of the Neanderthals. They often show considerable dental attrition due to the hardness of the food (45), as well as deposits of tartar.

Broca has distinguished several degrees of dental attrition ranging from zero to a complete wearing down so that the pulp cavity is exposed; nevertheless even people with teeth which are that worn down seem to have had little or no inconvenience from it. Grit, coming from querns or grain pounders, is the worst offender. If the average is known for a certain population group, the degree of attrition may be an important help in determining approximate ages. Pyorrhoea and abscesses of the dental roots occur in the late Palaeolithic, the Neolithic and a lot in Ancient Egypt. Pyorrhoea and arthritis are often regarded as being connected. Both affections were found on a skeleton of a Neanderthal from La-Chapelle-aux-Saints (161). Yet we have no proof of their connection. Besides, pyorrhoea does not necessarily develop more easily in people with an arthritic constitution, nor does an inflamed tooth necessarily lead to an arthritic condition. Pyorrhoea, unlike caries, is not linked with civilization. It is a microbe infection closely linked to a process of wearing down.

Peri-radicular osteitis is rare in prehistoric men. It does exist though, and Choquet (135) ascribes it to pyorrhoea. Perhaps it is due to abscesses caused by infection of the pulp cavity, exposed through dental attrition.

Alveolar abscesses are quite frequent and according to Moodie they already occur at the beginning of the Tertiary. He mentions a case in Hyracodon, a primitive rhinoceros from South Dakota dating from the Oligocene, where the abscess had penetrated to the mandibular canal. Pales (135) thinks the walls of the cavities of the lesion are too smooth to be caused by an abscess and rather suggest a cyst. I already have mentioned an alveolar abscess with osteitis of the whole mandible in one of the skeletons from the burial cave of Furfooz in Belgium.

Alveolar abscesses with fistulae occur in fossilized animals. They are rare in men and only one case is known: a lower jaw of an adult from Krapina has two open fistulae on the outside of the bone.

Dental abscesses where the tooth comes out and the alveolar bone is absorbed, have been found in skeletons from the shell heaps of Algeria, dating from the Aurignacian; also the Neolithic period, from Ancient Egypt, and Pre-Columbian America.

Lesions caused by sinusitis maxillaris seem to occur often in the Neolithic period. Four cases among Pre-Columbian inhabitants of Peru have been described by Moodie, although with certain reservations because three of these also had a cranial tumour, and the other showed a healed cranial fracture.

Dental malformations have been described frequently. Hypercementosis is very common. This affection, also inaccurately called radicular exostosis, consists in the formation of new cement and is caused by inflammatory irritation. This irritation originates from infection of the alveolus. If it is slight but prolonged, hypertrophic cementitis arises which tries to hold back the infectious process. If this cementitis is localized a periapical club is formed. If diffuse the roots will become rugged and also increase in volume considerably. Hypercementosis is not necessarily linked to caries; for this reason we have the name primary hypercementosis. Yet caries may result in hypercementosis in which case it is secondary.

Split roots of the front teeth occur particularly in the Palaeolithics of Krapina and the Neolithics of the Vendée. This phenomenon is sometimes taken to be an ethnic feature characteristic of dolichocephali, and sometimes considered an expression of civilization (149). Fusion of the roots of the molars also occurs in the skeletons of Krapina.

Hypoplasia of the enamel occurs in all periods, and the fossilized human remains of Grimaldi shows a particular fold. Often Pre-Columbians have a cavity on the inside of the incisors which results in shovel-shaped teeth. This characteristic does not occur only in these people.

Absence of the lateral incisors also occurs frequently. It is difficult to make out whether it is congenital or a development disturbance of the tooth. H. Brabant (16) describes three cases of agenesia of the premolars in skeletons from Spiennes, dating from the Neolithic. Baudouin (11) finds two cases of inclusion of the canines in the remains of bones from Vaudancourt, an anomaly which seldom occurs in the Neolithic period. A deviation which consists of the persistence of the milk canines has been found in a Pre-Columbian skull from Peru and has been described by MacCurdy (113) as follows:

In front of each first premolar is a small alveolus proving that the milk canines were still in situ at decease. Their persistence caused the malplacement of both permanent canines; the left, twisted on its axis, is seen directly in front of the lateral incisor; the right did not erupt at all, its impacted position being shown in the radiograph.

Erosion is common and is, according to Siffre, proof of developmental disturbances in prehistoric men. It is the expression of profound pathological disturbances during life, which have however been overcome by the sufferer.

While discussing dental anomalies we might add Carabelli's cusp, first described in 1844. It always occurs on the facies palatina of the first upper molar (Pl. 42). It makes this tooth into a sort of fifth cuspid, though it is not part of the chewing surface. On the contrary, it is situated on the side of the tooth, slightly above the neck, and it hangs down. It can occur either on the left or the right side, but usually on both. It is found in 50 per cent of the Neolithics.

Rouillon (149) concludes from statistics that it is characteristic of dolichocephali of small stature. It has been fiercely disputed as a feature of heredo-syphilis but now it is accepted simply as an archaic condition.

MacCurdy mentions a large osseous cyst of the maxillary central incisor at the right in a skull from Pre-Columbian Peru. The same skull has two trepanation apertures.

In collaboration with P. Wernert, Pales (134) describes a tumour of the lower jaw of a bovida found in Achenheim (Lower Rhine). It is a large cavity, causing the loss of the 1st and 2nd molars. The cavity extends backwards under the roof of the third molar and forwards under the third premolar. As a result of the tumour the inferior side of the lower jaw is swollen. Its wall is smooth and the cavity does not show any covering layer. The cavity and the absence of teeth have influenced the position of the other teeth so that the wearing down of these was abnormal and took place in step formation. Because of these facts Pales excludes inflammatory lesions and osseous tumour, as well as myeloplactic growth. The neoplastic central form of actinomycosis is discontinued in favour of a growth of cystic character, originating from the teeth themselves, causing the bone to swell but not to crack, and thrusting out the teeth. He regards it as a radiculo-dental cyst or a cystic adamantinoma. Although it is a benign tumour, in the course of growth it must have stricken the nerve and interfered with chewing, so that the animal experienced repercussions in the form of under-nourishment, which might possibly have led to its death.

It is impossible to make out whether dental surgical therapy existed in prehistoric times. Tooth extraction may have been practised just as it is among primitives today. Neolithic dental lesions have been described as having been caused by toothpicks. If the Egyptians and Etruscans used gold, it certainly was not for therapeutic purposes, but rather for aesthetic reasons.

The origin of dental caries is still obscure. The first known case of caries dates from the Permian and has been described by Renault. It is found in mastodons and Pleistocene elephants. But it certainly does not occur among the Neanderthals. There are however a few cases which date back to the Aurignacian, but their origin is not certain: one is the so-called Aurignacian skeleton of Libos (France), which was found only slightly deeper than the Neolithic layers. A brachycephalic skull from Hungary, studied by Von Lenhossék does not allow reliable dating. Von Lenhossék thinks caries was introduced into Europe by brachycephalic invaders from Asia at the end of the Palaeolithic period. I cannot agree with this view: indeed, in the Iberian peninsula we see the Mesolithic develop directly from the Magdalenian culture – a skeleton from the Asturian with a pronounced dolichocephalic skull, which is trepanned, shows slight caries of the left third molar (25). The evolution of these Spanish prehistoric cultures does not point to a necessary change in the race or to influences and infiltrations from the East.

If Palaeolithic dental remains are scarce, so that it is dangerous to draw far-reaching conclusions, we certainly can say dental caries is a disease of civilization. Dental caries occurs with certainty only at the end of the Palaeolithic period, when domestication not only promoted brachycephalization but also caries. The same can be said of Egypt: caries rarely occurs during the pre-dynastic period, but increases with development of civilization and more among upper class people than among ordinary people, who ate uncooked and hard food. When in the Ptolemaic and Byzantine periods civilization reaches the lower classes, caries occurs more among them as well. Caries was wide-spread also among Pre-Columbians in America. Drawing a statistical line which reflects a true picture is difficult because each investigator premises his own conditions. Mummery, studying English material comes to the following figures:

|  | *per cent* |
|---|---|
| 68 Neolithic dolichocephalic skulls | 2 = 2·94 |
| 32 Bronze-Age skulls | 7 = 21·87 |
| 59 Early Yorkshire-dolichocephalic skulls | 24 = 40·67 |
| 44 old mixed skulls | 9 = 20·45 |
| 143 Roman Period skulls | 41 = 28·67 |
| 76 Anglo-Saxon skulls | 12 = 15·78 |

Fifty years later von Lenhossék drew up similar statistics for central Europe, based on about a thousand skulls. He assumed that each tooth lost during life was a tooth with caries, a fact which Mummery had not taken into account. He arrived at the following very high figures:

|  | *per cent* |
|---|---|
| First century B.C. | 85 |
| Fourth century A.D. | 83 |
| Thirteenth century | 86 |
| Recent skulls | 90 |

French Neolithic skulls show dental caries in 3 to 4 per cent of the cases. This corresponds to Mummery's figures for England.

Elliot Smith has studied five hundred skulls of aristocrats from the Old Kingdom, buried near the Giza pyramids and he came to the conclusion that formations of tartar, caries, and alveolar abscesses occurred as often as in Europe today.

From all this we can conclude that dental caries occurs in Europe only from the Neolithic and Mesolithic periods onward. It is true that the teeth of the skull from Broken Hill (Rhodesia) show bad caries, although the skull has been ascribed to the Neanderthal (15). Nevertheless the origin of this skull is doubtful.

Several factors have been mentioned in connection with caries: rough food, repeated action of traumata, influences from the surroundings, ethnic factors, and others. Rough food, possibly mixed with grains of sand, can hardly be taken to be a damaging factor, for sand leaves clear marks on the teeth. Moreover, the badly worn down teeth of Palaeolithic men, where attrition sometimes reaches the pulp cavity, show no traces of caries. Natives of New Caledonia, who eat very rough food, have practically no dental caries, and are regarded as the race with the strongest teeth.

Dental caries seems to be connected with arthritic lesions in some races. Moreover, Neolithic men have dental caries usually localized at the level of the neck of the tooth. This type can easily be compared to that which affects arthritic people. Pales recognizes no connection between the two affections: osteoarthritic lesions have been found in large numbers in creatures from reptiles onward to Palaeolithic men, although no dental caries was to be traced. The same applies, to a lesser degree, to the natives of New Caledonia.

Dental caries occurs more in granity regions than in chalky regions, although the chemical composition of the teeth is the same. The surroundings seem to have some influence. Moodie assumes that dental caries originates from the acidifying of nutritional remains in between the teeth. A. Åslander does not agree with this view (4). He assumes that dental caries is the result of two factors: deficiency of minerals and sugar. He prevented dental decay in his children by administering minerals in the form of osseous flour. In his view the alkaline quality of the soil, due to abundant rain and lower evaporation, causes a shortage of minerals in food. In contrast to more southern regions this ends up mainly in the sea. Eating herring with bones and all, as our ancestors did, is the reason that they never suffered dental caries. (Åslander makes up for this deficiency with osseous flour rather than with calcium components, because the first also contains indispensable trace-elements.) The other factor is sugar, though he does not assume that fermentation is the decisive cause, but rather the combination of glucose with calcium to form calcium saccharate. This is possible with 50 per cent solutions of glucose, a condition which is fulfilled when sucking a lump of sugar. The pH is not important, for fermentation could never reach a sufficiently low pH, and certain food, such as lemon juice, apples and other fruit, are much sourer. The contact does not last long enough. During fermentation not enough H ions are produced to bring the pH down. Åslander also points out the protective part played by saliva and the 'plaque'. He also excludes bacteria since the sugar concentration is not optimal for development of their pathogenic qualities. He concludes that sweets in particular are very cariogenic. Fluorine alone has no absolute value in preventing caries. On the other hand refined sugar is more harmful than chewing the simple sugar-cane, since the latter also contains mineral elements. And, according to Lødrup, a healthy set of teeth can be kept with a low fluorine content if strontium and vanadium are present to a higher degree. Åslander himself found caries among people who drank water with a high fluorine content in Sweden.

Hereditary and racial tendencies towards caries seem greater among dolichocephali than brachycephali. Nevertheless the Neolithic people from England were less subject to caries than their French neighbours, who were made up of at least two races and three or four ethnic groups. These factors, considered on their own, do not give a satisfactory explanation.

Diet seems to be of prime importance. Leigh made an interesting study of this in America: comparing the degrees of dental caries among several Indian tribes, he came to the conclusion that caries was rarer in tribes who lived exclusively by hunting, and more frequent among tribes which applied themselves to agriculture and in particular to cultivating maize. In other words, civilization brings with it caries. The same can be seen in people who become civilized and start eating the white man's food: sugar, soured or fermented drinks, over-refined prepared food. This may also explain why caries occurs in general more in dolichocephali than in brachycephali: the former seem to have lived at a more cultivated level than the latter. At any rate, the explanation of artificial cranial deformation is based on this (Pl. 43): the long skull being the symbol of a free man (Delisle).

Without wishing to go into details about the question of artificial cranial deformation, it is to be noticed that MacCurdy (113) found two types in Pre-Columbian Peruvians: the first being the coastal type, the second the mountain, or Aymara type. The first consists of a fronto-occipital flattening, the second of a circular constriction with compensatory elongation. Excavating in the Highlands he found 341 skulls of which 147 were deformed. The deformation occurred more among women than among men: of 133 male skulls 58 were deformed (43 per cent); of 107 female skulls 64 were deformed (60 per cent). Of 47 trepanned skulls he found 21 skulls with the Aymara type deformation.

The higher level of living of dolichocephali is the reason for their teeth succumbing to caries first.

CHAPTER 20

# Tuberculosis

It should be made clear that every case of tuberculosis of the bones consists even more than non-specific osteomyelitis in secondary metastasis. This starts from a primary tubercular focus, established through an aerogenous infection – seldom by an enterogenous one – and with interference from a lymphatic gland, an haematigenous metastasis may result, which in fact causes a miliary form of the disease. This may not always be noticeable because its course depends on a series of factors which themselves depend on the resistance and reaction of the body as well as on the virulence of the bacilli. The dissemination of bacilli in tissues may cause a tubercular focus in a particular organ. We are interested in localization in the skeleton in particular.

In the bone itself tubercular bacilli are usually localized in the growth centres and more in the spongy part than in the compact part. In young people they are localized in particular in the epiphysis and metaphysis, in adults in the bones with red marrow, such as vertebrae, phalanges, metacarpals and calcaneus.

Although many anatomo-pathological pieces have been found dating from before the Pleistocene, and although some authors speak of tuberculosis in dinosaurs, others do not accept this diagnosis, which after all is only based on a certain number of vertebral ankyloses which could just as well – perhaps with more justification – be ascribed to osteoarthritis (124). No find in animals so far justifies the diagnosis of tuberculosis.

Le Baron (105) describes three cases in Neolithic skeletons which he diagnoses as coxalgia; the actual pieces have been lost, however. He also mentions a case of tubercular arthritis of a right clavicle from the cave of L'Homme-Mort (Lozère). A similar lesion is described by Houzé, in his study of the Neolithic remains of Sclaignaux (Belgium):

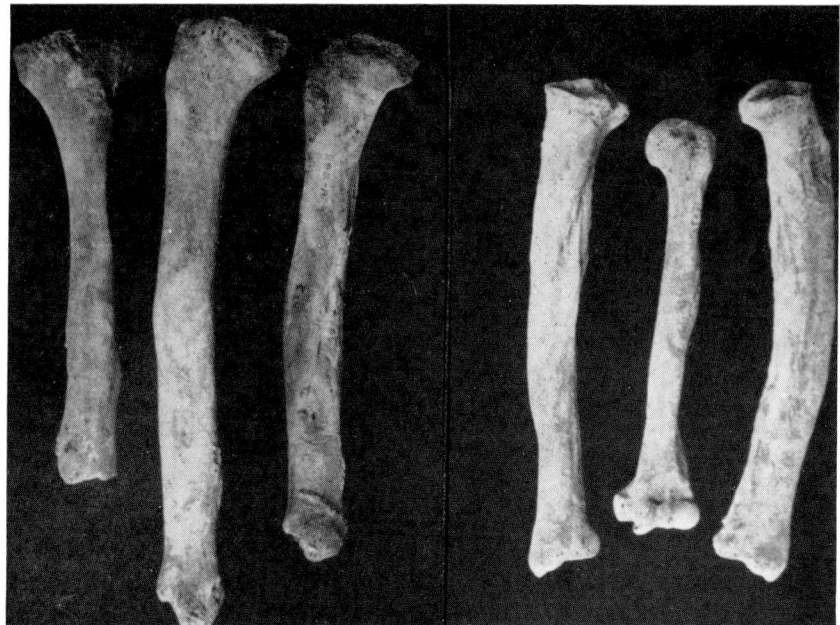

46. (left) three tibiae with syphilitic lesions, from Paucarcancha. (right) two tibiae and one humerus with identical lesions, from Patallacta. After MacCurdy (113).

47. Pre-Columbian skull with syphilitic lesions. After Mac-Curdy (113).

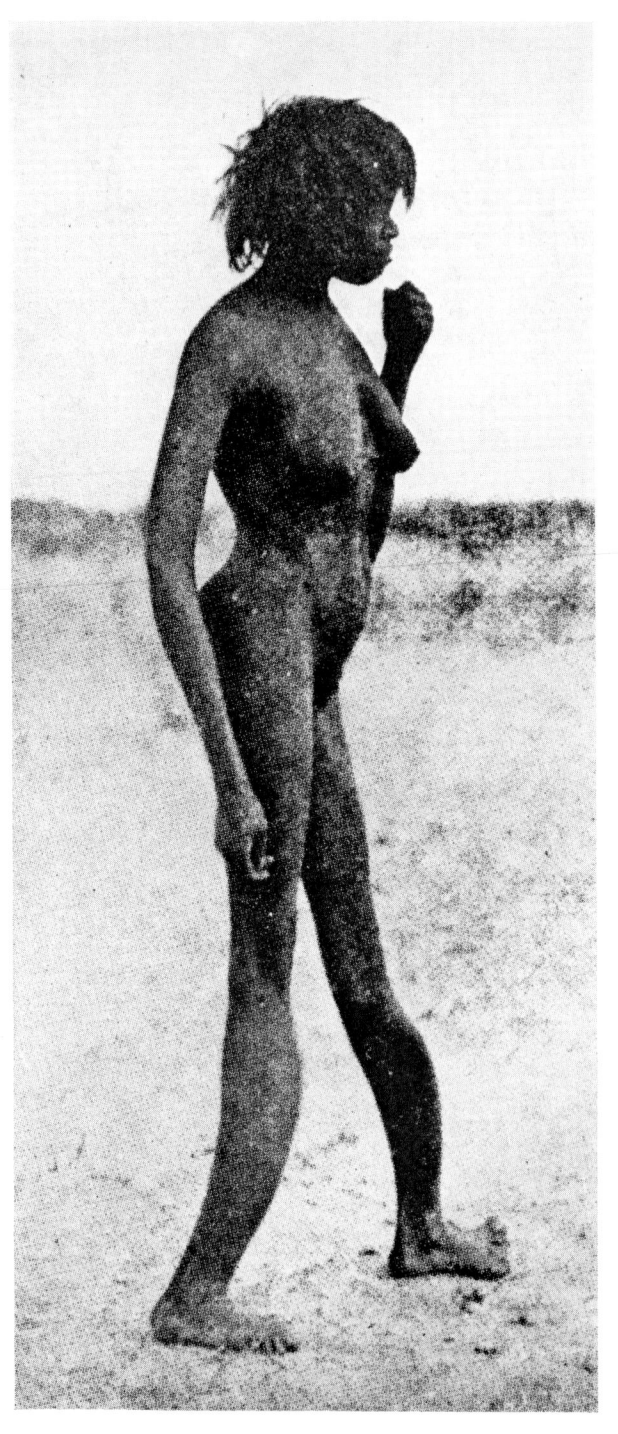

48. Tibia *en lame de sabre*, or 'Boomerang leg', in a woman of forty, from Central Australia. Ciba symposium, 1940, No. 2.

49. Stele of the Priest Ruma. From Wells, *Bones, Bodies and Disease*.

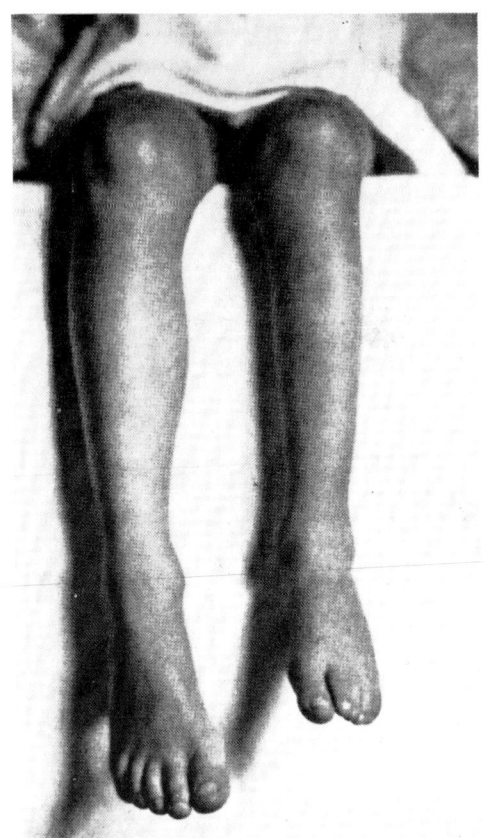

50. Photograph of a lesion in a boy of eight, the result of poliomyelitis in early childhood. From *Triangel*, II, No. 7, 1956.

A fragment of the right clavicle of an adult shows, on the inferior surface near its distended sternal end, a cavity surrounded by a flange of exostosis which is very probably the result of osteoperiostitis. The position and form of the lesion make a tuberculous origin probable for this affection.

Marc and André Romieu (135) have also described vertebral lesions in skeletons from the Bronze Age, which mainly took the form of a hollow with one-sided collapse of the vertebra and scoliosis. Although they had reservations about the possibility of Kümmell-Verneuil disease, Pales definitely regards it as tuberculosis.

Dr Baudouin described three Neolithic metatarsal bones, affected by osteitis together with cavity formation, which suggest healed spina ventosa.

Pales has drawn attention to an anatomo-pathological specimen in the Musée de Paris, which shows a peculiar ankylosis of the tibio-talo-calcaneus joints. The piece has also been described by Broca. The lesion must certainly have festered for a long time: a scar from necrosis or caries is noticeable in the tibio-talar joint. Broca (22) observes that these lesions may heal spontaneously through months of resting, that is, he adds, if the patient is helped by his fellow tribes-men. This means that even if no real medical help existed, this circumstance points to the existence of a nursing help. This anatomo-pathological specimen is said to come from a dolmen from the Lozère but this is not absolutely certain. We do find one particular characteristic of osseous tuberculosis: we have already said that osseous tuberculosis affects the small bones and the extremities of the long bones in particular. The spreading of the inflammation to the nearby joint results in chronic arthritis, attended with a swelling of the joint to a knob, which is called the tumor albus.

I have already mentioned the possible tubercular vertebral lesions in Bronze Age skeletons. As this localization generally takes an important place in the study of palaeopathology, it seems appropriate to point out that the disease was described for the first time in 1779 by Pott. In 1816 the French surgeon Delpech made clear its tubercular origin.

The differential diagnosis of tubercular spondylitis on isolated vertebrae is not easy. The relatively slow evolution of the disease gives rise to the different appearances the lesions may have in the anatomo-pathological pieces. As in all forms of osseous inflammation there are stages of demolition of the bone and stages of recovery, and in chronic cases both may occur together. When the lesion heals, more or less pronounced changes may take place in the spinal structure. A clinical examination may reveal other symptoms

H

which will lead the differential diagnosis in one direction or the other but we lack these when examining single vertebrae. Moreover a radiograph of the initial stage is quite characteristic because of the changes on the intervertebral disc which is apparent in a reduction of this cavity. It is difficult to work out whether a destructive process in an excavated bone is due to physical conditions in the soil or to pathological causes. For the same reason it is difficult to differentiate spondylitis, originating from bacterial diseases. Finally I should like to point out that syphilitic gummata, lymphogranulomatosis, Scheuerman's disease, osteitis fibrosa generalisata, certain metastases, multiple myelosis, diseases of the lymphatic system, haemangioma, and certain tumours should be taken into account in differential diagnoses.

An example of a possibly healed process is the skeleton of a young man, found in Heidelberg and dating from the Neolithic period. It was found by P. Bartels (9) with completely destroyed 3rd, 4th, 5th and 6th dorsal vertebrae, and a resulting kyphosis (Pl. 44). Yet it cannot be said with certainty that this is a case of healed Pott's disease. Williams thought it was perhaps an impacted fracture.

From the anatomo-pathological pieces which have been found up to now we can say that we have no proof of the existence of tuberculosis in the Palaeolithic period, and that it probably does occur from the Bronze Age onwards.

Lung tuberculosis has not been found in Egyptian mummies. This does not necessarily mean that the disease did not occur. Osseous tuberculosis occurred a lot in Egypt: of this we have certain proof. The Egyptian burial ritual is the cause of the seeming absence of lung tuberculosis. Though the mummies have been preserved, the lungs have always been removed. Neither can the bacilli be detected in mummies because they perish quickly after death of the host. Nevertheless the mummy of a priest of Ammon from the XXIst Dynasty (1100 B.C.) from Thebes furnishes unmistakable evidence: the first lumbar vertebra and the four last dorsal vertebrae have been destroyed. Kyphosis has resulted from this as well as the classical psoas-abscess (Pl. 45).

That the 'white death' occurred frequently is shown by lesions of Pott's disease in skeletons from a grave of ten individuals from the predynastic period. A figurine found in the desert in the vicinity of Assuan, dating from the same period, probably pictures someone suffering from Pott's disease.

Pre-Columbian material almost never shows lesions which might indicate tuberculosis. One has to bear in mind of course that this material is very difficult to date. Many Post-Columbian cases are known where tubercular lesions do occur. For this reason the generally accepted opinion is that tuberculosis did not occur in America before Columbus's arrival and was brought there by the Europeans. The discovery of America has had far reaching results in pathology, even more if we accept that syphilis was brought to Europe from there: one half of the world exchanged a social pest with the other half.

Attempts to clarify the problem of the occurrence of tuberculosis in Egypt as well as in America have been undertaken, through the study of pictures and figurines.

Pictures of hunchbacked figurines are in many cases thought to portray Pott's disease. It is difficult to accept that hunchbacks evoke such characteristic forms that the aetiology can be recognized from a picture or a figurine. A distinction should be made between strongly angled hunchbacks and regular kyphotic ones. The first groups are usually due to trauma, to several forms of spondylitis, and con-genital wedge-shaped spine (which is rare) (42). Regularly kyphotic hunchbacks may result from Scheuermann's disease, which itself is difficult to differentiate from spondylitis ankylosans. Besides this some forms of fracture, multiple spondylitis foci, senile osteoporosis, osteomalacia and Kahler's disease may cause this deviation. To be complete scolioses should also be mentioned. Angular scolioses may be the result of spondylitis or inborn malformations. An idiopathic form of scoliosis also exists, but it can also be found in Scheuermann's disease and osteoporosis. In other cases thoracic scoliosis is the result of retraction processes, or it may be com-pensatory in lumbar scoliosis; which itself is caused by an affection of the hipjoint, a shortening of the lower limb, a one-sided paralysis from poliomyelitis or spondylolisthesis.

Besides the hunchbacked figurines we have further evidence of the existence of tubercular spondylitis in a Mexican figurine with a presternal tumour 'which may be a cold abscess of the sterno-clavicular joint'. It is my opinion that this should be regarded rather as perforating aortic aneurysm. Besides, this affection seems to have occurred more frequently than a tubercular one on the bones in this particular anatomical area. The existence of aneurysm has been ascer-tained on certain skulls because of the enlargement, brought on through aneurysm of the Arteria carotis interna in the cranial basis of the carotia.

In his conclusion Pales (135) makes the rather risky remark that if we can ascertain that tuberculosis and dental caries only occur since the Neolithic period, then even though no direct connection exists between the two, yet they may still work together, since tuberculosis is a demineralizing factor for the teeth, which are then affected by a secondary infection: dental caries. I cannot agree with this opinion because for instance pregnancy is certainly a condition in which the demineralizing factor for the teeth is even greater than tuberculosis, and in spite of this we find no dental caries in the Palaeolithic period.

# CHAPTER 21

# Syphilis

Whether syphilis occurred in pre- and protohistoric times or not is a question beset by even more difficulties than the study of tuberculosis in the same periods. Indeed, attributing certain lesions found on prehistoric material to syphilis means making a stand against those who hold that syphilis was brought into Europe by Columbus's crew via Spain in 1493, and via Naples in 1495 (170). The latter base their opinion on the fact that this disease was discussed in the Diet of Worms in March 1495, where it was regarded as a fresh sign of divine vengeance. In August of the same year an imperial edict was published ordering the town councils to expel from the community all sufferers from syphilis. A similar document, which is the oldest known written document mentioning the occurrence of the disease in France, dates from April 1496 and is kept in the communal archives of Besançon (register C.C.55). It mentions among other things: 'En cette même année, avril, dix personnes atteintes de la maladie, dite de Naples, expulsées, reçoivent chacune un florin ou dix gros'. From Naples the disease seems to have been spread throughout the whole of Europe by Charles VIII's returning soldiers. The first soldiers from the Naples expedition were back in Lyons at the end of October 1495, the last ones by Spring 1496. Already on 7 July of the same year the 'Veyrolliers' are subject of a complaint of the *Consulat* which was handed over to the Duke of Orleans, Lieutenant of the King (41). We know the first occurrence of the disease in England through the edict of the 'Council of the borough of Aberdeen' of 21 April 1497, which attracted the population's attention to the disease and gave the best way to fight it.

In his well-known poem written in 1535, Fracastorius thinks that syphilis, also called the 'French disease', came into being simultaneously in America and Southern Europe.

Another difficulty, possibly the greatest, lies in the fact that syphilis causes lesions which show a close similarity to those caused by the tropical disease, framboesia or yaws.

There are three theories concerning the origin of the disease: the first theory says that syphilis has existed since time out of mind, spread all over the world, and that the fifteenth-century epidemic was only a flare up of the existing disease, activated under influence of the American virus, brought to Europe by Columbus's sailors. The second theory says that syphilis did not exist in the eastern half of the world before the discovery of America. This leads directly to the third theory, which simply says that it is impossible to say what caused syphilis because the second theory accepts syphilis as a fact in Pre-Columbian material, which is not at all certain.

Dujardin (43) points out that the fact that a disease has not been described in Antiquity does not mean that it did not exist: as an example he cites scabies, a disease which has not been described but which certainly existed. Syphilis has been overlooked because syphilitic forms then were not so malignant as in the sixteenth century. He finds a number of descriptions by Romans, Greeks, Jews, Hindus, and especially by Chinese, which correspond to certain forms of the disease. He cites Hoang-Ty (2637 B.C.), who describes a chancre, from which the virus spreads into the bloodstream. A treatment of rubbing with mercury compounds is advised.

It is striking however that descriptions of these affections never mention any connection with bone lesions. Furthermore, no signs of syphilis have been found in Egyptian mummies, nor could Ruffer detect any on the bones of the Greek soldiers of Alexander, buried near Alexandria.

The first cases of syphilis among Columbus's sailors were described as 'bubas'. Historians have added that this affection was contracted from Red Indian women, who experienced no great inconvenience from it and probably had always suffered from it (La Casas).

Testimony to the transfer of syphilis from America was given by Ferdinand Columbus, Christopher's son, Diaz de Isla, Oviedo y Valdes and Las Casas. Ferdinand Columbus writes that when his father returned to the Spanish peninsula only 160 Spaniards of the contingent that had sojourned there remained, and that all of them were affected by the 'French disease'.

Ruy Diaz de Isla has given the most convincing proof of the way the disease penetrated our continent: it appeared in Barcelona in 1493 and had been brought by 'the crew who had had intercourse with women of the [discovered] land, where the disease has existed since time immemorial'.

Oviedo speaks in the same way: 'Many times in Italy did I laugh when the Italians named it the "French disease" whilst the French called it the "disease of Naples". Both sides would have uttered truth had they called it the "Malady of the Indies".'

The great difficulty lies in interpreting the material correctly. Syphilis can affect the bones and when this happens the lesions can sometimes be so characteristic as to exclude all doubt. But besides these, there are a great number of lesions which are not so characteristic. Even today it is difficult to diagnose with certainty a number of deviations which are suspected to be of syphilitic origin. In some cases even the sero-diagnostic method falls short.

D'Harcourt (40) gives the following description of a skull and bones which he regards as bearing syphilitic lesions:

The affected region comprises the frontal bone, both parietals, the occipital and the mastoid processes. The surface of the right parietal shows elevations, which are fairly variable though weak, about a centimetre across and between which there are depressions. Below the parietal eminence two very small perforations are seen, no doubt attributable to the disease. The anterior half of the occipital is affected. In the middle of this surface a rectangular depression ($12 \cdot 5 \times 20$ mm.) produced by active ulceration, can be seen. On the anterior half of the left parietal one sees only an uneven surface but the other half is grossly affected. There one finds, below the parietal eminence, an area looking like ivory which measures $30 \times 40$ mm. Below this, on the inside, are four perforations: three small, which are not due to post-mortem erosion, and one which is larger, being 28 mm. in its greatest length and 10 mm. in greatest width, which cannot be attributed to either an injury or a trephination. Lower down there are five small areas of well marked ulceration. The internal table has a smooth, unbroken surface. The sutures have fused prematurely. Some teeth remain in the upper jaw but the mandible is missing. The skull is that of a man aged about fifty years.

The most interesting long bones include the two femora and the left tibia. The femora present some thickened and glossy areas which are evidently due to periostitis. One of the tibiae is the most affected piece. The absence of sequestra and cavitation in the long bones increases the probability of this being a case of syphilis. The bones have been radiographed and, in addition, two of them have been sawn lengthwise. The findings from simple external examination are fully confirmed by the other two methods. The narrowing of the medullary cavity is very marked in the tibia. The osteoperiostitis of the Paracas long bones is very probably syphilitic in origin.

MacCurdy (113) also describes a few similar cases in Pre-Columbian Peruvians (Pl. 46).

We shall here consider the anatomo-pathological lesions which syphilis causes in bones: contracted osseous syphilis from the second-ary stage onward causes lesions which mainly affect the skull and tibiae. These only consist of inflammation of the marrow and dis-appear without leaving any trace. Only at the end of the second stage and in the course of the third do more serious lesions develop, namely syphilitic periostitis and osteitis and periostitis gummosa.

Syphilitic periostitis shows no specific characteristic in itself. Only the localization and osseous malformation which it causes give a distinctive character to the lesions. Histologically there is a picture of banal subacute periostitis. The periosteum is infiltrated by lymphocytes and plasmocytes, proliferates abundantly, and forms new connective tissue, which becomes the site of important bone forma-tion. These lesions are usually situated near the skin: on the skull, the protruding part of the collar bone, and in particular on the crest of the tibia. The latter has a thickened appearance, in particular seen from the front: the bone bends and takes on the shape of the scabbard of a sabre (150).

Gummatous lesions can develop as central gummata in the bone itself or as periosteal gummata on the periosteum. The first form gelatinous masses, which destroy the bone and proliferate. The nearby periosteum reacts through formation of extensive exostoses. The gummata can proliferate or absorb: often they perforate the bone and exude a syrup-like fluid through a fistula. Periosteal gummata tend to affect the bone, in which they eat deep holes and furrows. In the skull destruction is sometimes so thorough that the dura mater is bared. The skull assumes a 'worm-eaten' appearance. Around the ulceration an intense osteoblastic reaction sets in through which a protruding hyperostotic wall of bone is formed. Super-infections play an important part: their action is associated with the primary lesion and gives rise to far-reaching sequestra-tions. Destruction of the nasal bones and cranial bones should be attributed to these secondary infections rather than to syphilis itself. Here one should be very careful of possible confusion with 'goundou' (framboesia).

This description is best illustrated by a case, mentioned by MacCurdy (113), and very similar to one described by D'Harcourt. It is a skull of a 26-year-old man from Pre-Columbian Peru. The author writes:

51. Negative drawing of a hand, above the picture of a horse. Pech-Merle.
From Breuil, *Quatre cents siècles de l'art pariétal.*

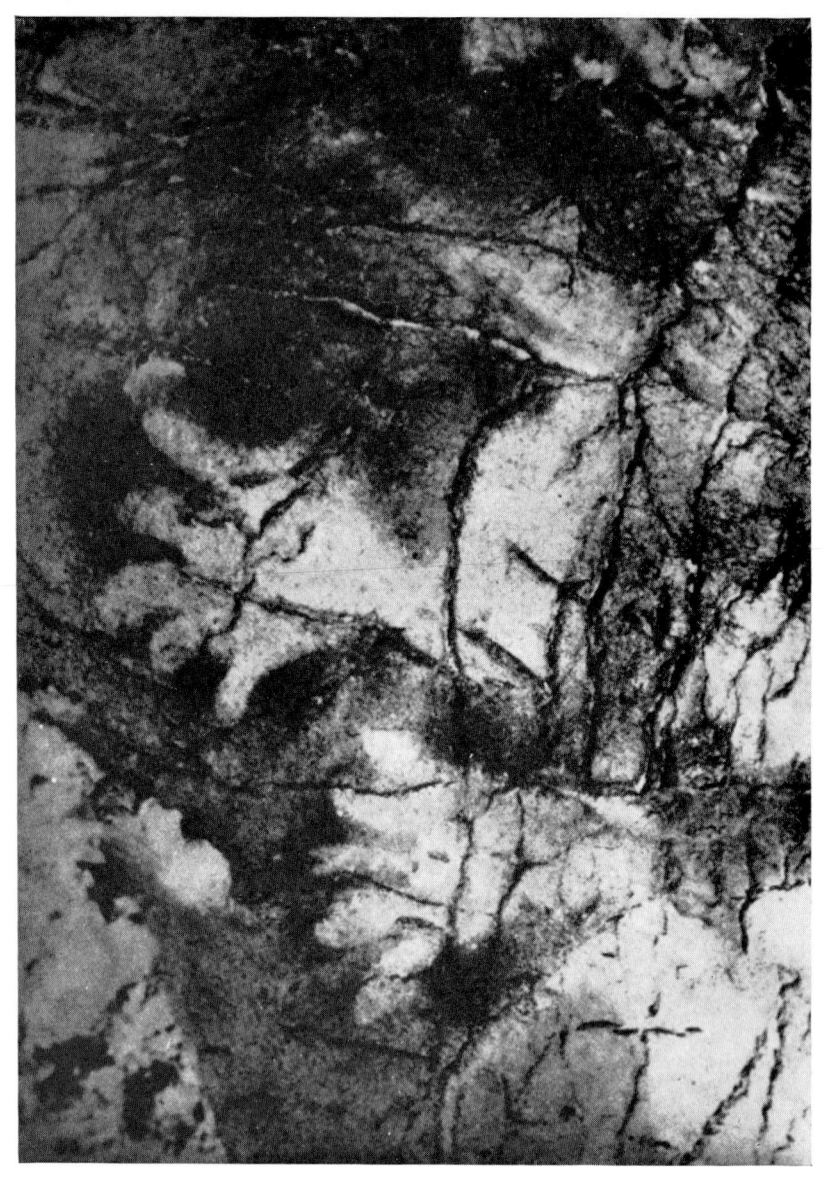

52. Representations of mutilated hands. Cave of Gargas. After Breuil, *Quatre cents siècles de l'art pariétal.*

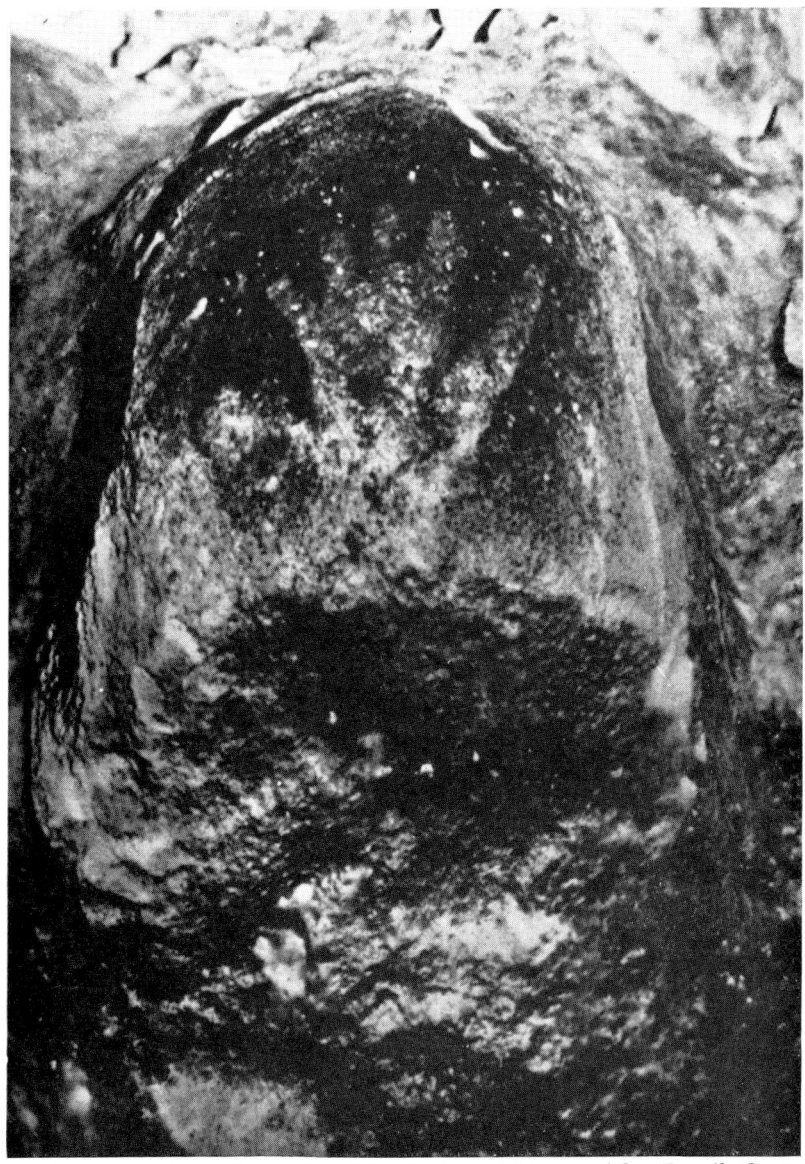

53. Representations of a mutilated hand. Cave of Gargas. After Breuil, *Quatre cents siècles de l'art pariétal.*

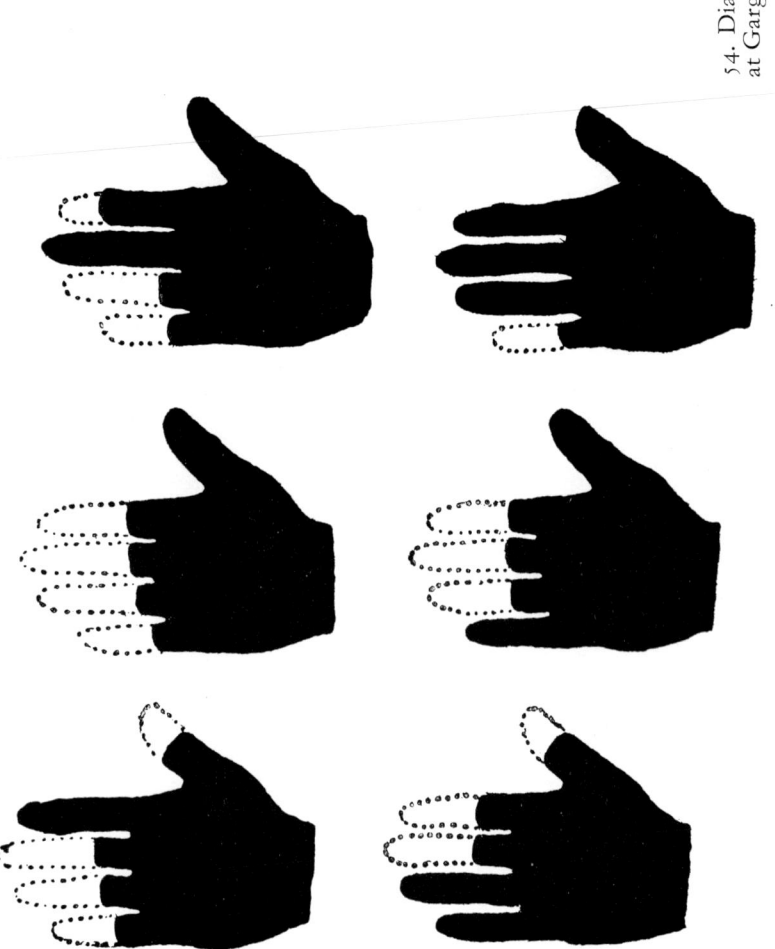

54. Diagram of the most frequent mutilations at Gargas. After Casteret, *Dix ans sous terre.*

The disease had eaten away the external table over a circular area 4·8 cm. in diameter, two-thirds of which is in the right parietal and one-third in the frontal, and had even penetrated the diploë and internal table near the centre of the affected area, leaving a hole some 8 mm. in diameter. Two contiguous fields in the region of the obelion are also affected to the extent of complete removal of the external table. Just back of the left stephanion is an oval cicatrice 2·3 by 1·6 cm. in dimensions, apparently due also to disease. In the anterior temporal regions there is a suggestion of a former hydrocephalic condition. (Pl. 47.)

Besides bone destruction syphilis leads to a considerable new formation of bone: it is at the same time destructive and constructive, which is its essential difference from tuberculosis.

From the above description we can see how difficult it is to differentiate a specific infection from a non-specific one, in particular when there is secondary infection.

Congenital syphilis can cause the same lesions as acquired syphilis, but usually expresses itself in the form of osteo-chondritis or Parrot's disease. This form is usually to be seen in new-born or still-born heredo-syphilitic babies. The modifications are most easily seen radiologically. They consist of a change of the epiphysial line. In normal circumstances this line is straight and thin. In a pathological condition it is widened and appears in the form of a yellowish band with protruding angles and notches and is best compared to an electrocardiographic tracing.

In view of these facts it is immediately clear that it is impossible to trace this form of syphilis, because it affects in particular the cartilage, which is not preserved.

The material said to show syphilitic characteristics is limited. A skeleton found at Solutré by Abbé Ducrost in 1872 shows pathological tibiae with extensive exostoses. The left tibia shows one situated in the middle of the diaphysis, the right has three spread over the whole diaphysis. The age of the skeleton is doubtful; but if it does not belong to the people of Solutré the latest it can date from is the Roman period, in which case it certainly would be a proof of the existence of Pre-Columbian syphilis in Europe. But the presence of exostoses does not itself prove this to have been a case of syphilis. If this were so, it would be remarkable that no other such lesions have been found. Rollet has an explanation: 'This tribe [the people of Solutré] had been living in such isolation that it died out without transferring the disease to other tribes and peoples.' This assertion is based on Lortet's theory that the people belonging to the Solutréan culture did not produce a lasting line of descent, to mix with later peoples, and that the bearers of this culture died out when the reindeer and horse herds disappeared, which were the mainstay of their existence. These people came from Asia, from where they brought the disease; with them too the disease also disappeared.

But in the cave of Parpallo (Spain) we can ascertain a degeneration of this fine Solutrean culture into one of the Perigordian type, called by Paricot-Garcia as the final Solutreo-Aurignacean. This fact disproves the theory of Lartet.

Pales (135) discusses another series of bones which have been described as being affected by syphilis, and refutes the arguments used to prove syphilis in this material. Among others there is a femur, which he describes as affected by Paget's disease. Parts of the parietal bones of a child, however, diagnosed as heredo-syphilis, had been misrepresented by H. U. Williams as a case of symmetric osteoporosis.

In England, probably due to the influence of James Hunt, who denies the occurrence of syphilis in prehistoric times, no lesions have been described as syphilitic.

In Egypt several bones have been found with lesions which have been ascribed to syphilis. The most important of these is the skull from Roda, near Karnak, described by Lortet in 1907: it shows serpentine ulceration of the left parietal, which caused actual perforations of the bone, and irregular white spots, as a result of degeneration of the tabula externa in several places. There were also similar erosions of the occipital in the top portion, the left frontal end of the eyebrows. Nowhere are there exostoses to be found. Nor is there any trace of bone reaction. Lortet thought the patient, a woman of about twenty, must have died so quickly that the exostoses did not get a chance to develop. Gangolphe has come to the conclusion that the lesions should rather be attributed to post-mortem action by rodents and beetles.

K. Jäger (74) describes bones with syphilitic lesions from the ossuary of Chammünster, Greding and Ardenback dating from the early Middle Ages. He mentions nine cases of hyperostosis luetica (eight tibial and one femoral), osteosclerosis, eburnation, and gummata of the skull. He found only four femora with rickets, which does not seem very likely to me.

In America things seem clearer (180): the discoveries made by Joseph Jones in 1876 in prehistoric layers of Tennessee seem certainly to be affected with syphilis. Yet even against this evidence objections have been raised over the difficulties which exist in distinguishing specific from non-specific infections. Even more important is the fact that after the arrival of Columbus the natives carried on burying their dead in the same manner as they had always done (Wolff). Thus the difficulty in dating still remains. Dating by the Carbon-14 method will in future throw more light on the problem than any of the minute descriptions of the pieces suspected of syphilis. Here also, as in Egypt, there is the difficulty of distinguishing syphilis from the tropical disease, framboesia. As far as America is concerned some have suggested the possibility of leprosy as the cause of the described lesions. The facts do not seem to bear this out. It may be true for Europe; indeed Broca and Raymond have found, in examining leper burial places, many bones with lesions which suggest syphilis, quite apart from the fact that the latter is convinced of the existence of syphilis in pre-history. He bases his assumption on a few pieces from Baye's collection. The number of cases of leprosy diminished considerably as soon as mercury treatment came into use; this is another factor contributing to the confusion.

Not all the observed symptoms on the above-mentioned bones have a specific character, such as exostoses, cranial ulceration, synostosis of the cranial fissures, hyperostosis, tibia *en lame de sabre*, deviations of the roof of the mouth, dental deviations, such as hypercementosis and enamel hypoplasia. Moodie (124) warns us about lesions which have been accepted as irrefutable signs of syphilis. The tibia *en lame de sabre* is not necessarily linked to a luetic factor. This phenomenon, also named platycnemia, can be a morphological characteristic and useless on its own in diagnosing syphilis if the other symptoms are absent (Pl. 48). The same applies to Hutchinson's teeth and Carabelli's cusp, inherited from man's predecessors (100).

If the few pieces which show so-called syphilitic characteristics are the only material to convince us of the existence of syphilis in prehistoric times and if we consider that the examination of more than 25,000 Egyptian skulls proved to be negative for syphilis, this all seems contradictory to the very character of the disease: syphilis is a social plague, against which no race or class is protected. It becomes epidemic only during certain periods and in countries where no efficient treatment of the disease, or prophylaxis, exists. Once it has taken root it is not to be stamped out. It travels with the armies from one country to the other. Furthermore syphilis has a tropical character. Egypt certainly would have been a centre from where the disease could have spread into Europe. It becomes even more complicated when we realize that in our regions rickets causes difficulties with the differential diagnosis of syphilis: here is the reason why Parrot regarded so many symptoms of rickets and dental erosion as syphilitic. For this reason we must be sceptical when reading the description of possible syphilitic lesions on the bones from the ossuary of Chammünster, where these lesions surpass in number those of rickets. Pales, Williams, and Jeanselme have challenged Parrot's diagnoses, when reviewing the material. Ruffer, Elliot Smith, Wood-Jones and Moodie are among those who deny the existence of prehistoric syphilis.

Perhaps C. H. Hackett's suggestion (191) that treponemata of more benign forms such as pinta led through mutation to the eventually more malignant stage of treponema pallidum, may satisfy both sides.

# CHAPTER 22

# Poliomyelitis

Did infantile paralysis exist in the remote past? 'And Jonathan, Saul's son, had a son that was lame of his feet. He was five years old when the tidings came of Saul and Jonathan out of Jezraël, and his nurse took him up, and fled: and it came to pass, as she made haste to flee, that he fell, and became lame, and his name was Mephiboseth.' (II SAMUEL iv. 4).

Paralysis after a fall practically never occurs among children. Yet usually parents see a seeming connection between a fall and the chance following of poliomyelitis, so that a reading of the Book of Samuel compels us to consider possible poliomyelitis.

The stele of an Egyptian priest, Ruma, from the XVIIIth dynasty (Pl. 49) (c. 1500 B.C.) shows him with the right leg atrophied to a high degree (70, 100, 161). The leg is shorter and a pes equinovarus has developed. As mentioned above, the Pharaoh Siptah had a similar deviation. Fanconi (48) speaks here of *Missbildung*, which he ascribes to a paralysis resulting from poliomyelitis suffered during youth and resulting in a failure in bone development. He bases his assumption on comparison with modern cases (Pl. 50).

In the seventh century Paulus of Aegina (625–90) described a 'colic disease' occurring as an epidemic, leaving paralysis of the limbs, which regressed in the course of months. If these are not cases of poliomyelitis, the virus of the disease certainly is closely connected to it.

In the eighteenth and nineteenth centuries Underwood (1784) in England and Monteggia (1813) in Italy described a 'disease, not yet described, which is characterized by a period of a few days of fever and results in paralysis of the legs in the period of recovery'.

A skeleton from the predynastic period dating from about 3700 B.C., found by Flinders Petrie in Desbasheh (south of Cairo), shows a shortening of 8·2 cm. of the left femur. A stick had been buried together with the body.

G. H. Rolleston described an English Neolithic skeleton from Cissbury as having signs of poliomyelitis, because the left humerus was 3·25 cm. shorter, and the left radius 2 cm. shorter than the corresponding bones of the right arm.

All these facts may point to poliomyelitis, but often diseases of the nervous system can cause similar deviations, such as cerebral infantile paralysis, where the brain may even be affected before birth. Ætiological factors are mainly encephalitis and brain trauma. Other factors are: alcoholism of the parents, hereditary predisposition (so that many think that trauma only causes lesions when a predisposed basis exists), malformation of the brain, and all diseases which lead to imbecility. The aetiological factors do not give a constant or sound picture of the disease. Thus hemiplegia spastica infantilis, which arises very early in childhood and is caused by encephalitis, is described in this group. Usually it follows after convulsions. In this picture the first mentioned case fits best. Another picture is that of Little's disease, a summary of many cases of diplegia and paraplegia in as far as they all have birth trauma as an aetiological factor.

American physicians ascribe it to haemorrhages destroying the cortex of the brain, while Van Gehuchte sees the cause in an insufficient development of the pyramis cerebri, also as result of birth trauma. Other cerebral diplegiae and paraplegiae complete this series.

Since palaeopathology should tell us whether or not disease is invariable, and if new diseases have come into being, it should also solve the problem of whether poliomyelitis is a new disease or not and, if it did exist in prehistoric times, whether its clinical appearance has become more frequent.

A conclusive answer to the first question cannot be given, because methods of investigation available do not allow us to find out with certainty whether the disease existed in antiquity or not. The second question, the more frequent occurrence of the disease, or in other words whether poliomyelitis is a 'civilization disease', has been answered affirmatively by Fanconi. According to him several factors work together. First there are the extreme measures of hygiene taken in the supplying of drinking water and the drainage of waste matter, which diminished the possibility of fecal-oral infection, which in fact builds up a natural immunity. This may be the reason why older individuals, in particular, become victims: they do not possess this natural immunity and suffer poliomyelitis with very bad paralysis symptoms. Increasing social intercourse causes an extreme mixture of several viruses. From statistics of infant mortality in Switzerland Fanconi proves that countries with a high rate of infant mortality will have fewer cases of poliomyelitis, because many children die very young, even before poliomyelitis has been diagnosed. If in these countries children become infected, there will be less chance of becoming ill because of their natural immunity. In more civilized countries, with lower infant mortality rates, the symptoms come more to the fore because the conditions are the opposite.

# CHAPTER 23

# Other specific infections

We have little information about other specific infections. Indeed often they appeared as epidemics and during epidemics embalming was impossible because of the high death rate. In the Bible several are described, but the written word alone does not always allow us to recognize a disease with certainty, not even the writings of so brilliant an observer as Thucydides.

## PLAGUE

I have already mentioned in an earlier chapter that Ruffer described a possible case of plague, basing his assumption on a microscopic study of the bacilli present in mummies. Yet he did not find any typical 'buboes' in embalmed remains. One point of interest is that plague did not occur in England before the Crusades. Indeed only then was the rat brought to the island, where, like the rabbit, it did not previously exist (191).

## SMALLPOX

Another mummy from the XXth dynasty shows a skin covered with a peculiar vesicular eruption. This rash looks strikingly like small-pox (161).

## LEPROSY

A case of leprosy has also been described. This affection must have been widespread, for it was already known in early historical times – indeed, Elephantiasis graecorum, known by the Ancient Greek doctors of the Near East, corresponds symptomatologically with leprosy. George Sticker, however, points out that lepers and

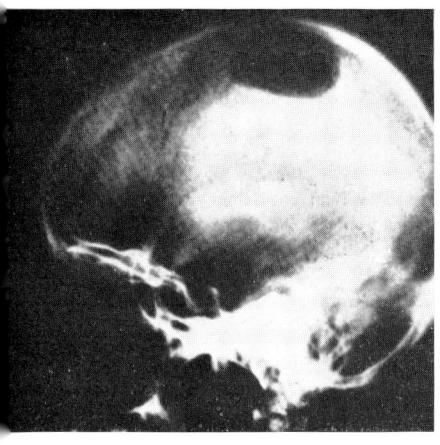

55. Congenital opening of the os parietale (Prof. Wertheimer). *La Presse médicale.*

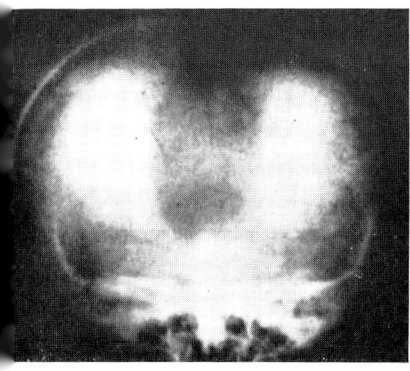

56. Skull with a case of dysostosis cleido-cranialis (Brailsford). *La Presse médicale.*

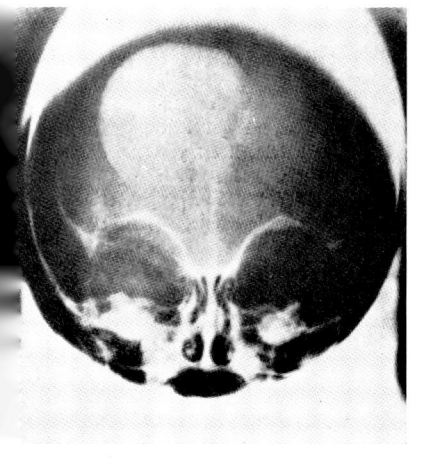

57. Meningocoele (Prof. Wertheimer). *La Presse médicale.*

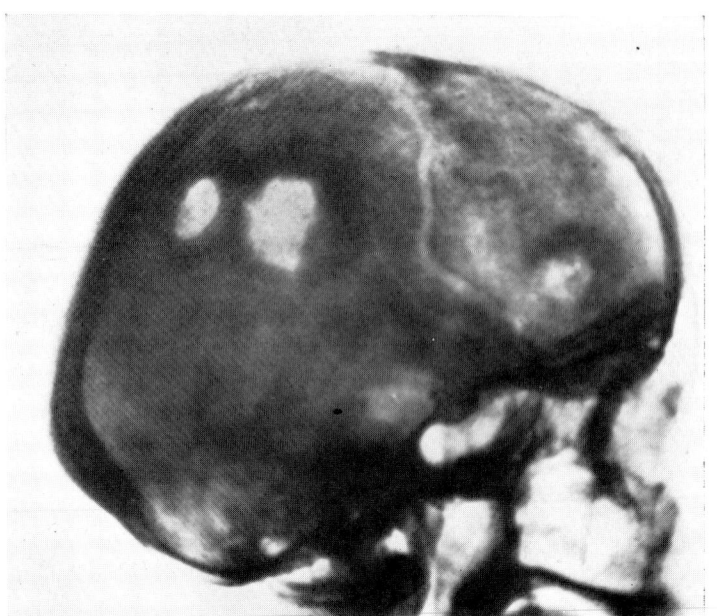

58. Tuberculosis of the cranial roof (Pr. agr. Lecuire). *La Presse médicale.*

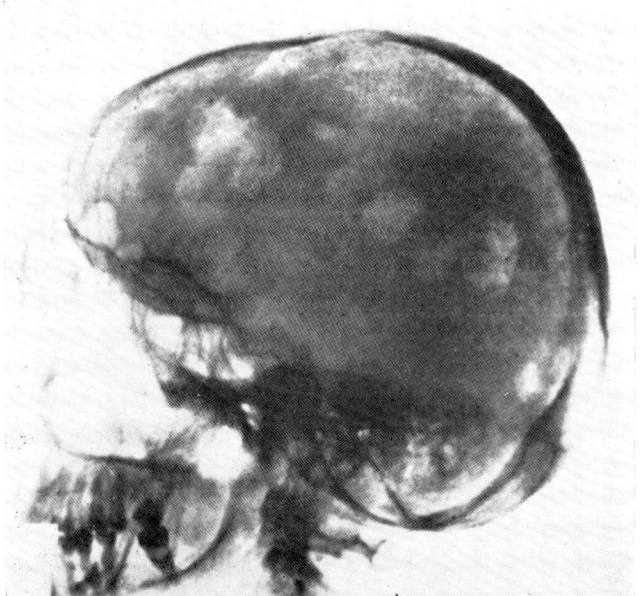

59. Syphilis of the cranial roof (Prof. Wertheimer). *La Presse médicale.*

victims of syphilis have always been excluded from the community and were not given a proper burial. The remains of their bones have therefore been less protected against disturbing influences, which explains, according to him, the rare occurrence of material affected with either leprosy or syphilis. In lower Egypt elephantiasis graecorum certainly was widespread. According to Galen the affection is seldom found in Germans, but more in Celts, in particular among the Gauls (70). The possibility of leprosy should be borne in mind when interpreting the prehistoric cave paintings of hands with mutilated fingers, as in Gargas. I shall devote a special chapter to this matter.

## MALARIA

Even though we have so far found no direct proof of the existence of malaria, we can safely assume its occurrence from the many cases of enlarged spleen found in mummies. This enlargement is indeed the normal expression of the disease.

## SLEEPING SICKNESS

This also concerns nagana. Tsetse flies (*Glossina*) have been found in Oligocene strata in Colorado (124). The possibility of the existence of sleeping sickness in ungulates during the early Tertiary (about one million years ago) is not excluded, although Trypanosoma cannot be traced and the disease does not occur in the Colorado region today. The disease does not have any bone lesions.

## ACTINOMYCOSIS

The specific infection caused by the actinomycosis fungus has been described earlier on as occurring in cave bears. Moodie (124) describes a case in a rhinoceros, genus *Aphelops*, from the middle Miocene.

In prehistoric Peru we find a few other specific infections, not so much in anatomo-pathological specimens, but more in figurines and decorated jugs. The depiction of the lesions as suppurative 'lumb-jaw' is typical.

## GOUNDOU

In the same way we can infer that in prehistoric Peru goundou must have existed, although it no longer occurs there. The disappearance

I

of the disease allows us to mention again the possibility of the existence of sleeping sickness in Colorado.

## VERRUGA PERUANA

Another disease often depicted is verruga peruana, or Carrion's disease, an affection which is more malignant than smallpox. This disease killed one quarter of Francisco Pizarro's army. Two forms are distinguished: a serious one, with a wartlike rash, high fever, bad pains in limbs and head, which results in death; and a less malignant one with fever as in malaria, and the familiar eruption. The papilla sometimes causes the formation of real growths (40).

## UTA

Uta, an ulcerous disease, has also been depicted in prehistoric South America. Until recently the disease had been regarded as a prehistoric form of syphilis and leprosy. Now we know it is leishmaniasis. The disease begins with the formation of a small papilla, which grows gradually. After a few weeks to two months an ulcer sized 1 to 3 cm. has formed. The characteristic places for infections are the face, mouth, lips, ear and neck. This disease is often confused with blastomycosis, which causes very similar lesions.

## NIGUA

Finally I mention nigua, a sort of sand flea. We find small figurines showing men busy removing the eggs of this parasite from the soles of their feet.

# CHAPTER 24

# Diseases of the soft tissues

Palaeopathology has so far provided little or no information about diseases of the soft tissues. I shall therefore devote attention to Egyptian mummies, which allow us to study the soft organs as far back as 4000 B.C. This period corresponds to the Neolithic period and the beginning of the Metal Age in our regions. There are no direct historical connections between the two regions but as has already been mentioned in the introduction, in a somewhat later period commercial relations may have existed. This contact may have brought with it notions of medical treatment.

The following cases have been compiled by Sigerist (161).

Although to us arteriosclerosis might seem bound up with the strains of modern life, it has been found frequently in Egyptian mummies. The influence of alcohol, tobacco, and syphilis does not seem of primary importance to the aetiology of the disease. Meneptah's mummy – according to the Bible he was drowned (Exodus xiv. 27, 28) – shows typical arteriosclerotic lesions and dental deviations which point to a less dramatic death from old age. Even now the lesions can be examined microscopically in those parts of the aorta or in arteries of limbs which have been left in the body by the embalmers. The lesions consist of atheromata, atheromatous patches, and ulcers, more or less calcified. Ruffer put the pieces first in a solution of sodium bicarbonate 1 per cent and formaldehyde 0·5 per cent for two days. He then placed them in alcohol treated with saltpetre. Then they were washed and treated as modern preparations.

Arteriosclerosis plays an important part in cerebral haemorrhages, followed by paralysis. Facial paralysis is one detectable form, though the cause is not necessarily arteriosclerosis. A Tlingit helmet from British Columbia bears the picture of an old man with left facial paralysis (191).

Though the lungs in mummies are shrivelled, we can still detect anthracosis in an individual who lived during the XXth dynasty. In another mummy of the same period we find symptoms of pneumonia in the stage of hepatization, caused by the presence of many cells with nuclei in the lung alveoli. A mummy of the Greek period is known in which inflamed patches were found in the upper parts, while similar lower ones had healed. In the liver and lungs bacilli were discovered, which look like those which cause plague. Yet no buboes were found, so that a certain diagnosis was not possible.

Pleurisy has often been observed through the adhesions caused by the disease.

A mummy from the XVIIIth to the XXth dynasty has been described with congenital atrophy of the right kidney, while another mummy showed multiple abscesses on both kidneys. Three bladder stones have been found in a skeleton from the pre-dynastic period. Three cases of gallstones are also known, though Moodie found only one such case. He is surprised at the rare occurrence of lithiasis in Egypt, in contrast to its frequent occurrence now. I believe that stones in kidneys, bladder, and gall must have attracted the attention of the embalmers. Indeed, magical powers have always been attributed to these stones. It seems sound to assume that the embalmers searched for them to sell them as magical objects.

Calcified eggs of Schistosoma haematobium have been noted in the kidneys of two mummies from the XXth dynasty. This parasitosis is widespread in North Africa. In connection with the problem of cancer it is peculiar that lesions of the bladder can give rise to neoplastic degeneration. This phenomenon has not been noted in South Africa, where this parasitosis occurs. The material however is too sparse to ascertain these facts in antiquity. We can only say that the disease already existed 3,000 years ago. The main symptom is haematuria. This phenomenon has been mentioned in Egyptian papyri, but this does not necessarily indicate bilharziasis, as the causes of haematuria are various.

Though the Egyptian mummies are shrivelled and dried out, they still can give a vivid picture of most moving tragedies. The young princess Hehenit of the XIth dynasty shows a vesico-vaginal fistula, probably caused by a protracted delivery, as a result of her abnormally narrow pelvis. This delivery also caused her death. A Coptic negress suffered the same fate during delivery and was buried with the baby's head still lodged in the pelvis. Her pelvis also was too narrow. She was also crippled as a result of a missing sacro-iliac joint and had been raped. This fact indirectly became the cause of her death.

A young Nubian girl, while pregnant, suffered the most terrible cruelty. She had been beaten so that forearms and feet were fractured. She finally succumbed to a cranial fracture resulting from the beating.

Gastro-intestinal affections are seldom mentioned, mainly because these organs are not easily preserved, though adhesion and chronic appendicitis have been noted. Gallstones in a mummy from the XXIst dynasty is the only known case, as mentioned. Intestinal and vaginal prolapsus on the other hand occur quite regularly.

A case of gout in an early Christian is perhaps worth mentioning (see chapter on arthritis).

We should take into account that prolapsus viscerum, found in mummies, is not always of pathological origin. It can have been brought on *post mortem* by the process of desiccation. In fact it is a similar process to that which takes place in older people *intra vitam*. There the cause is a shrinking of the soft tissue in the pelvis.

Specific infections of the soft organs are described separately. Tuberculosis and syphilis are the subject of separate chapters. Besides these, cases are now and then described which can have side-effects in the skeleton.

Eye lesions were very common in Egypt, as they still are today. Here, however, the major factor seems to be a religious consideration: the will of Allah. I can personally testify to the fact that until recently in Cairo countless numbers of flies swarmed round the eyelids of Arabs affected by trachoma. The fact that one tenth of the recovered prescriptions of Ancient Egypt refer to these affections proves that eye lesions were a frequent occurrence (69).

Although we have had no equivalent to Egyptian mummies to test for affections of the soft tissue for the corresponding periods in Europe, I should like to attract the reader's attention to two sources which are able to provide more information. First there are the extraordinary preservation conditions in the peat soil of Scandinavia for proto-historical times, where fleshy parts of the body are well preserved and secondly there are the prehistoric pictures on cave walls in which engraved and painted forms sometimes suggest certain deviations. In discussing the Venus figurines we have already studied one of these facets more closely.

# CHAPTER 25

# Medical views on prehistoric representations of human hands

Besides classical mural paintings, representations of human hands have been found in certain caves (75). Often these are represented in connection with an animal figure (Pl. 51). In other cases they occur separately, without any apparent connection with an animal figure, as if they had been painted with no definite purpose. So far about fifteen caves with pictures of human hands have been discovered. The most impressive are the caves of Gargas, Bédeilhac, Trois-Frères and Cabrerets in France, Altamira, and Castillo in Spain (18).

We can regard these representations as one of the first forms of pictorial art; they are even considered as the very first stage. They were made in the Aurignacian period (140). We shall call them 'positive' pictures when the hand has been covered in paint and then pressed upon the cave wall, and 'negative' ones when the hand has been held against the wall and then sprayed with paint so that the broad outline becomes clear. This technique might surprise us because we associate it with action-painting, a technique regarded as very modern. Almost always they are the outline of the left hand, from which we can deduce that prehistoric man was right-handed. In the Cave of Castillo twenty-three impressions of left hands have been found for only one of the right hand. One can also assume from this that left-handedness at any rate existed in those times. According to Breuil, 124 representations of left hands and only thirteen of right hands occur in the Cave of Gargas. These figures differ by a fair margin from those quoted by E. H. Verbrugge (173), who mentions that he found 114 pictures of hands with only three of the right hand.

Of special importance to palaeopathology is the fact that in the Cave of Gargas some images show one or more missing fingers. Sometimes even all fingers and thumb are missing, so that only a stump is left. Until recently these mutilated hands had only been found in the Cave of Gargas (Pls. 52, 53). In 1956 J. Jolfres (121) discovered six negative mutilated hand figures, painted in ochre in the cave of Tiberan. This cave is situated on the eastern side of Gouret mountain; on the western side lies the Cave of Gargas. In 1960 the *Bulletin de la Société préhistorique Française* announced that similar mutilations had been discovered in a Spanish cave (20). Here only the little finger was involved. E. H. Verbrugge, however, has given the following figures for yet another Spanish cave, the Cave of Castillo: forty-four representations of hands of which thirty-five are left ones. When visiting the cave I could not find this number of pictures. I could only ascertain that the pictures mainly occur on the right side of the cave passage, as one enters. The same author also mentions one case of mutilation in Castillo. This I was also unable to find. The guide, Felipe Puente, who is acquainted with all the existing pictures, categorically denies the existence of representations of mutilated hands in the Cave of Castillo.

The question concerning the significance of these hands and the causes of the mutilations rises spontaneously. A. J. P. Van den Broeck (171) is convinced that they are not mutilations. Indeed, he says, no mutilations have ever been found on 'positive' images. He assumes that the representation has a strictly personal character. which should be regarded as some sort of signature or 'name-card', whereby the maker, while painting, kept one or more fingers hidden by folding them in the palm of the hand. I cannot agree with this view because negative representations in other caves do not show these mutilations. Also it does not seem very logical that all painters should have presented the same 'name-card'. Besides, the outlines of the pictures are far too clear to have been made with a hand with folded fingers, as the distance between hand and wall would become too great and give a less well defined image. And even if it were possible, it certainly would not be so for the thumb since it is impossible to eliminate the distal segment alone through simply bending the proximal one. Rutot (153) is also unwilling to accept this opinion and would sooner accept the idea of mutilation.

Weinert and Casteret (28) think they are sacrifices. Casteret bases his hypothesis on a comparison with the customs of some primitives today, though in my opinion this method is dangerous and can lead to false conclusions. He cites many examples from the Pygmies and Hottentots. For the former such amputation is an expression of mourning or a means of obtaining a peaceful death, while for the latter it is a remedy against dangerous diseases. In this case however only the little finger is offered. During initiation rites of the Mandam Indians the forefinger and little finger of the left hand are amputated although the purpose is a mystery. Hoping to get rid of an enemy an Indian sometimes cuts off three fingers from the left hand, in order that his wish be granted. The natives of the Pacific carry out this mutilation on members of the family when they are ill or die. This amputation is performed piece by piece, day after day. In India a tribe is known as the Berula Kodo or 'finger choppers'. Every three years they chop off the little finger and ring-finger of a number of women during religious celebrations. No reason is known for this. In other cases the mutilation is carried out for practical reasons: some natives chop off the little finger of their wives to make it easier for them to make nets.

What is striking is that none of these primitives ever amputate the thumb. Some of the prehistoric men of Gargas however, had no thumbs. After having examined more than two hundred representations, I could not say that the amputations of any particular finger was preferred (Pl. 54). Nevertheless loss of the thumb does not occur as often as loss of any other finger. Besides, I think that intentional mutilation of the thumb would have been a most unwise act for primitives. Indeed the possibility of opposing the thumb against the other fingers, together of course with the development of the brain, can be regarded as a decisive stage in the development of civilization. Hands are for a primitive his most perfect tool, because of this possibility of opposition. Even in our age we still regard the thumb as vital: loss of a thumb in a work accident is regarded as a loss of almost 50 per cent of the usefulness of one's hand. Thus I am convinced that the riddle of the mutilated hands has an explanation in medical terms.

First of all we should consider the possibility of mutilation through an infection of the finger, or panaritium. Indeed it is logical to suppose that the primitive repeatedly received injuries when knapping flints and that the finger wounds stood a good chance of becoming infected through lack of hygiene and disinfectants. Van den Broeck (179) noticed that in prehistoric skeletons osseous deviations are never found on phalanges. Nevertheless the total of recovered phalanges is very small. We also know that when an infection of a finger penetrates to the phalanx, the latter usually festers until all the osseous tissue has been dissolved. But the fact that so many mutilations occur on one hand hardly allow us to ascribe them to panaritia. Another possibility is that the mutilations are connected with climatic conditions. It is certain that during the same glacial period there was a great difference between the Spanish and French climates. In Spain the climate was much milder than in France, where the mammoth lived, having disappeared from Spain by the beginning of the Aurignacian; indeed this animal occurs but once, in the cave paintings of Pindal (Spain). The climatic difference was due to the Gulf Stream, which influenced the Iberian Peninsula much more than the French meridional region. During this period the English channel did not exist and England was still linked to the continent. As a result the Gulf Stream flowed more westerly along the English coast. Dordogne and the region of Ariège suffered most, missing the warmth of the Gulf Stream and also being surrounded by high and cold mountain ranges, such as the Alps and Pyrenees. We might assume that the mutilations were the result of freezing since they occur more in representations which are regarded as belonging to children and young women, and this seems to be arguable since young people are more susceptible to cold. But this theory is no longer tenable since we now know that in Spain, which was warmer, figures of mutilated hands have also been found and that in France no other mutilations occur in caves near Gargas.

Gangrene is a condition which causes the same sort of mutilations as we are studying here. It is a result of interruption of the blood circulation. Many factors should be considered here, and again I emphasize the cold together with the fact that these mutilations usually occur in children, young people, and women.

Acute arteritis is a result of an infectious disease such as typhoid, paratyphoid, scarlet fever, smallpox, puerperal sepsis, appendicitis, dysentery, or of a suppuration near to an artery. These forms of arteritis are rare and have a gloomy prognosis, 50 per cent of the cases proving fatal. Less serious cases can lead to necrosis of a small member, such as a finger or a toe (128).

In cases of acute or sub-acute syphilitic arteritis the arteries at the base of the skull in particular are affected. In case of a chronic form, only parts of the arteries are affected, according to Darier. Grenet rejects this view. There is no agreement as to whether this disease manifests itself commonly in the peripheral blood vessels or not. I think however that this disease should hardly be taken into account since the prehistoric material, which should indicate the existence of this disease, is sparse and hardly convincing.

Arteriosclerosis could be relevant, but may be discounted since the representations usually point to young individuals. For the same reason embolic and diabetic gangrene may be eliminated. Important however are the so-called juvenile arteritides. The most interesting of these is Friedlander's presenile endarteritis obliterans. This disease is in no way distinguishable from the senile form, the only difference being that it occurs at the age of forty to fifty. Although the cause of the disease is not clear it is generally ascribed to excessive consumption of alcohol and tobacco.

Thrombo-angeitis obliterans, described in 1906 by Leo Buerger, is characterized by obstruction as a result of thrombosis, which affects the whole peripheral blood system, arteries as well as veins. This disease seems to occur most among male Jews aged between twenty and forty, and also Japanese and Englishmen seem to suffer from it. We may be certain that syphilis is not relevant to the aetiology of this disease. Besides, I do not think this specific disease should be considered in connection with Gargas.

Totally different is the situation regarding Raynaud's disease, which occurs in young women of about twenty-five. Bilateral forms are most common, although unilateral ones can also occur in the first stages of the disease. It further shows a hereditary character. Babies and young children can be affected. There are several aetiological factors: intoxication, acute or chronic infections, among which may be cited in particular syphilis, traumatisms, cardiopathy, such as mitral-stenosis, and also endocrine disturbances. The factors 'cold', 'young individuals', 'heredity' provide so much circumstantial evidence that we are led to assume that the disease is the cause of the mutilations of Gargas. True, syphilis, being a possible aetiological factor in Raynaud's disease, pleads against this supposition, but we do not consider it of first importance. Cold is not an absolute factor, as I have already said. It can only have had a secondary influence on a tribe, the victim of a hereditary disease, which on top of this must have accepted promiscuity as simple and natural, although this cannot be proved. Here then is an explanation why the mutilations occur only in Gargas and not in the other caves in the vicinity, where also representations of hands have been found.

Since I have been discussing the significance of the mutilations of hands in particular, it should be added that I do not reject the thesis that the representations of hands, including those which are not mutilated, be regarded as *ex-voto*. I do however strongly deny the fact that prehistoric man should have deliberately mutilated his hands for this reason. If he did so in Gargas, he would have done it in other caves as well, where unmutilated *ex-voto* hands prove the existence of an identical or at least analogous belief.

# CHAPTER 26

# Trepanation

Trepanation (91) is surely the most striking feat in the whole of palaeopathology. It has been performed since the end of the Neolithic period. In Spain, however, a skeleton from the Mesolithic period with a typical trepanation has been found (25) (see below). The existence of trepanation in prehistory was sufficient for Velpeau to declare: 'La chirurgie crânienne se perd dans la nuit des temps' (30). Although the first trepanned skull was found in 1685 by Mont-faucon in Cocherel (146) we have to wait till 1816 before Barbié du Bocage presented to the Société des Antiquaires a skull 'présentant un traumatisme qui avait fait perdre une partie du crâne, ce qui n'empêcha pas son possesseur de vivre encore de longues années'. Cuvier even thought that this skull, found at Nogent-les-Vierges, had healed so well that the patient survived the operation by about twelve years (6). Later in 1853 a similar skull was found at Crozon, but the discovery caused no excitement and it was completely passed over (31).

In 1873 M. Prunières discovered a part of a parietal with a small hole in a dolmen from Lozère. This fragment of bone probably had been worn as a pendant and used as an amulet. It was given the name of 'rondelle'. At that time he used the word 'trepanation' only to indicate the technique of making such a 'rondelle'; in the following year Broca gave 'trepanation' the meaning of some particular cranial treatment. In 1878 Lucas-Championnière still regarded trepanation exclusively as a ritual deed – but later he changed his mind and came to regard it as real surgical intervention, executed for decompressive purposes. Since then in France alone more than 200 trepanned skulls have been found. The fact that in the beginning they spoke of traumatism in connection with these skulls and only much later of trepanation, is the best way to show the scepticism which was felt about the possibility of such intervention in prehistoric times. And certainly holes in skulls are not necessarily the results of a trauma or trepanation (105, 139, 161).

According to P. Wertheimer, J. Avet, A. Levy, and J. Jenot (193) we can distinguish six types of cranial holes. The lesions resulting in the holes are not necessarily perforations. In radiology these are called 'lacunae'. They can be congenital, the result of trauma, of infectious or tumorous origin, of disease of the haematopoietic system, or of bone dysplasia.

Congenital holes of the parietals, described by Goldsmith in 1922 in skeletons which were completely normal, are generally only seated on the highest angle of the parietal (Pl. 55). The edges are smooth and clearly defined: the hole usually extends only a few millimetres but may in some cases reach a diameter of several centimetres. In Pierre-Marie's and Sainton's disease or cleido-cranial dysostosis, the lesions are symmetrical and always occur on the parietal (Pl. 56). Meningocoele or cerebral hernia cannot be confused with trepanation, because the edges of the bone are pressed upward and outward, which is not the case with trepanation where the edges remain horizontal (Pl. 57). To complete the survey I mention craniolacuna of the newly born, an affection which is connected with spina bifida, incomplete osteogenesis, craniostenosis or cranial tabes. These latter lesions are not necessarily perforations.

Traumatic lesions consist not only of trepanations but also of perforations made by forceps during obstetric intervention, which are not important in differential diagnosis of the subject under discussion. Only war wounds – in the broad sense of the word – and lesions caused by a fall on the skull are important here. These lesions are generally not circular unless they have been caused by a short, hard blow. Usually a stellate fracture is formed, the radii of which go out from the centre of the wound. If the blow is very hard, or if the wound is caused by a sling stone this symptom may be missing, but in that case the opening in the tabula interna will be larger than the one in the tabula externa.

Infectious processes may also cause perforating lesions. Such is the perforating osseous tuberculosis, described by Gangolphe, which causes circular holes in the skull, though here no osteophytes or hyperstosis are found on the notched edge. A ridge is formed around the opening on the outer face of the skull so that the tabula interna is more affected in fact than the tabula externa (Pl. 58). Syphilis too can perforate the skull. This can happen through necrosis of single gummata but usually the resulting lesions are multiple, as in Ludwig Pick's disease, which is characterized by passages with a 'worm-eaten' appearance (Pl. 59). If the lesion stands alone it is surrounded by a spiral groove, the result of an enlarged gummatous osteitis. Ordinary osteomyelitis and even myeosis can cause a perforation of the skull. Hydatid cyst should also be mentioned. Derry describes a skull of the Roman period from Egypt found in Shirafa; it shows an opening which looks like one caused by trepanation, and measures 24 mm. The lesion is irregularly circular and shows no sign of in-flammation. The edge is smooth and slopes outward 8 mm. from the edge. The bone is slightly lifted so that a flat ridge is formed. Below the opening, on the os occipitale, the bone is flattened. For this reason Derry ascribes the lesion to a dermoid cyst (Pl. 60). M. Kharadly (97) doubts the existence of trepanation in Ancient Egypt. I agree with this view because Egypt has left sufficient material, in the thousands of mummies.

Growths play an important part in the origin of holes in the skull, benign as well as malignant ones, primary or metastatic ones. The resulting lesion will even show the starting point of the process, as tumours of the exocranium will affect the tabula externa in par-ticular, while tumours of the cerebral membrane will rather damage the tabula interna. Tumours which should be considered are the primary benign ones, such as angioma, myeloplax growths, lipoma, fibroma, chondroma, and bone cysts; the primary malignant ones, such as bone sarcoma, fibrous sarcoma, chondrosarcoma and Ewing's tumour; finally metastases (Pl. 61) and tumours in the near vicinity of the skull, such as meningioma, glioma, cholesteatoma, angioma, as well as epithelioma of the sinus cavities and scalp. Tuberculous and syphilitic perforations in the skull are small, while tumorous ones are large. Perforations of the skull as a result of diseases of the haematopoietic system as well as those caused by bone dysplasia are less important. I draw the reader's attention to the fact that myeloma generally occurs in aged people and that the lesions are usually numerous, while chloroma and Ewing's tumour occur more in young people. Further, they perforate not only the cranial bones, but also the nasal bones and particularly the facial bones and the roof of the orbit.

As already mentioned we have to take into account old or recent post-mortem lesions, caused by chemical, physical and biological processes.

Among trepanations proper we should distinguish between those carried out on living persons and those performed after death. This distinction is easy for trepanations which show symptoms of healing. For the other cases it is less simple: a trepanation may have been performed after death, and the only criterion we have is the absence of healing reaction in the bone. On the other hand it may also have been carried out on a living patient, who survived the surgical operation for only a short time, so that the healing mechanism did not get enough time to leave clear marks. For this reason I think it is safer to consider them simply as trepanations with signs of healing and those without. In the second group we can also include the 'rondelles'.

A trepanation opening is usually larger at the level of the tabula externa than at the level of the tabula interna. This causes a sloping edge or lip (Pl. 62). The cells of the diploë are open and stay like that if the trepanation was performed not long before or after death. If the healing mechanism of the bone has had a chance to work, certain changes occur which are distinguishable with the naked eye or through radiographic examination. The edges of the opening are covered with a layer of compact bone tissue which is continued in the cranial bone. The appearance of this layer is totally different from that obtained from polishing. Post-mortem trepanations generally do not show a lip. The edges are uneven and show fine grooves which have been caused by cutting four grooves so that a quadrangular fragment of bone was obtained.

A radiological examination allows us to find out whether an opening in a skull is the result of trepanation or not. Guiard (66) distinguished three zones in trepanned neolithic skulls (Pl. 63). The first zone is a light shadow, a few millimetres wide. Then follows a more dense zone, and around this, at about 7 cm. from the opening, there is a rarefaction zone with dotted appearance. The rarefaction zone has also been described in modern trepanations by Mathey-Cornat and Guglielmi. The picture evoked is more or less like a rosette (Pl. 64). According to Guiard these zones have a prognostic significance: if the edges of the opening are irregular the patient died a few weeks after the operation. If the picture shows a dense zone, death occurred several weeks later. In cases with a rarefaction zone survival lasted more than one year.

Guiard thinks that, if the width of the dense zone exceeds 1 cm., the opening is not the result of a trepanation but should be ascribed to trauma. Tumours also cause a similar densification zone. To this are added osteophytes on the edge of the opening. Small osteophytes may form after a trepanation where infection set in and lasted for a certain time. Yet infection after trepanation occurred very rarely. An extended osteitis of the cranial roof causes a lesion which looks '. . . comme une aire moelleuse, produite par la pression des doigts sur une surface avec la consistance de mastic frais. . . .'

Perforations due to trepanation can be found on any part of the skull. Usually they are situated on the left parietal or occipital bone, less often on the temporal or frontal bone. The sutures were not avoided, which proves a great technical ability, for as may be easily imagined, the patient had no chance of survival whatever if the sinus venosus was opened. Cartaillac was impressed in particular by the fact that the operation was performed on the hairy scalp. Roy Moodie (124) has described a Peruvian skull with a trepanation hole in the frontal sinus, but this is a case of pseudo-trepanation, as the hole is an enormous fistula, resulting from a pan-sinusitis (Pl. 65).

In many cases several trepanation openings are found on one single skull. The opening is circular or oval, rarely square. In case of an oval opening its longitudinal axis lies in the longitudinal axis of the skull. The size of the openings can vary greatly: some tre-panations are not more than small points, others are enormous. In general the longitudinal axis is 4 to 5 cm. long and the breadth 3 to 4 cm. In some cases the edges of two or three trepanations cut through one another.

MacCurdy (113) describes a case of a Pre-Columbian skull with five trepanation holes (Pl. 66) as follows:

We come next to what is manifestly the most remarkable case on record, an adult male of about sixty-five years from Patallacta. This man underwent as many as five trephining operations all of which penetrated to the cerebral membranes and which were apparently performed at various times. In only one of the five is there any distinct indication of infection. Did the surgeons of that time possess any effective means of combating septicaemia? It would seem so; especially in view of the fact that the Incas were on occasions successful embalmers of their dead. Recently Reutter has made an analysis of embalming substances from Peru. These were found to contain *Baume de Perou*, menthol, salt, tannin, alkaloids, *saponines* and undetermined resins. Like the ancient Egyptians and Car-thaginians therefore, the Incas made use of substances rich in *acide cin-namique* to embalm their dead. As these have excellent antiseptic properties, it is permissible to presume that the surgeons of that time have taken advantage of their therapeutic value in trephining operations.

All five of the apertures are nearly round and vary but little in size (about 3·2 cm. in diameter). Two of these are on the left side of the frontal and so close together that only a slender bridge of bone intervenes. Had the two operations been performed at the same time there would have been either no bridge at all, or else a more substantial one. The upper of the two is probably the later operation. The third aperture is midway between bregma and obelion, its center being a little to the right of the sagittal suture. The fourth and fifth openings are wholly within the right parietal, one being between the eminence and the obelion. This was probably the latest of the five operations and is the one that was followed by infection. The external table has been removed from two oval spots on this skull, one near the right frontal tuber and the other near the left parietal tuber. Whether these represent minor trephining operations it would be difficult to say. Even if they do not, the skull would still be secure in its title to rank first in its class. This skull, with no signs of local pathological conditions and with nothing to indicate a single case of fracture, is one of the weightiest known documents in favor of trepanation for troubles that do not have their seat in the bony framework of the head. It also shows that trepanation was performed on the forehead as well as those regions generally covered by the hair. The bone had thoroughly healed after each of the five operations.

If the edges of trepanation holes sometimes have a different geometrical shape, or if some are not as steep as others, this is due to different techniques. These can be studied on skulls with incomplete trepanations, probably due to the fact that the surgeon stopped his work when the patient died in the course of the operation. Trepanations were also performed on dogs and human corpses as tests for new techniques which it is supposed could perhaps later have been applied by prehistoric surgeons.

One technique consisted of scraping the cranial bone with a flint scraper. Broca (22) experimented with this technique in 1887: the trepanation, performed on a young dog, lasted $8\frac{1}{4}$ minutes; performed on a fully grown one it lasted one hour. In this way Broca found that the flint instrument never damaged the dura mater. This process also explains the lip or sloping edge. If the long duration of the operation was a disadvantage, the advantage of this technique was that bone powder could be gathered together. Indeed until the last century this substance was used as a magical and pharmaceutic remedy.

A similar technique consists in the use of a polishing stone with rough grain. This causes the lip to be wider and the loss of tissue at the level of the tabula externa to be greater. Schmitt mentions that this technique was used for the trepanations in the skulls from the ossuary of Cornembeaux at Congy.

Lucas-Championnière (40) thought the method employed then was to drill several small holes in the skull, after which the bridges between them were sawn through one by one. Only one example of this has been recovered in a Peruvian mummy (Pl. 67). This technique would of course leave marks from the drilling holes on the 'rondelles', which in fact we do not find (36).

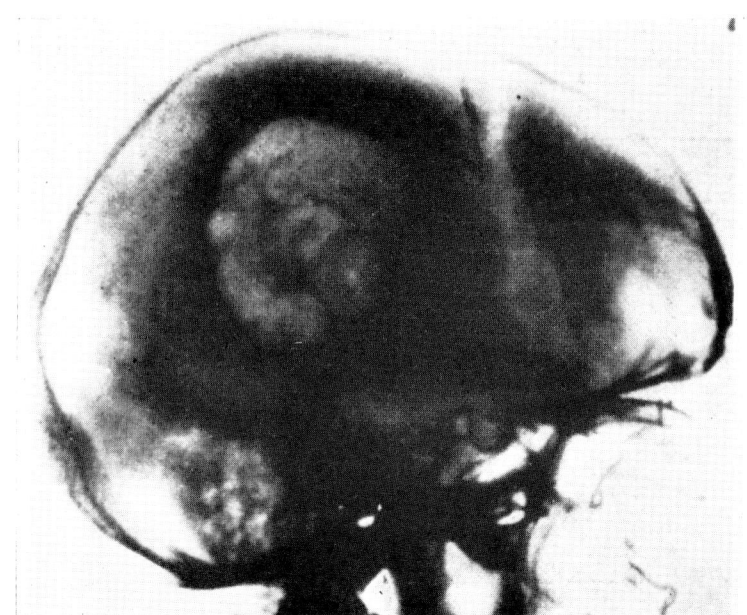

60. Dermoid cyst (Pr. Fontaine). *La Presse médicale.*

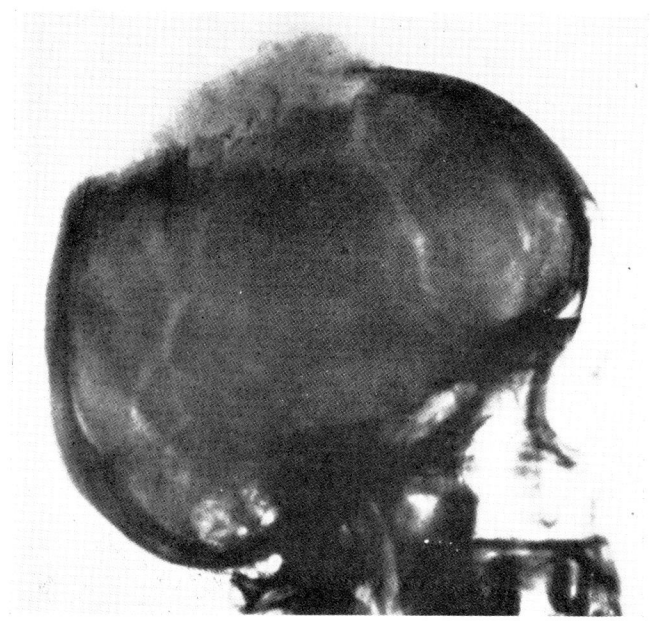

61. Metastases of a cancer of the thyroid gland (Prof. Wertheimer). *La Presse médicale.*

62. Healed trepanation opening. Saint Urnel. 1922–IV–12. P. R. Giot.

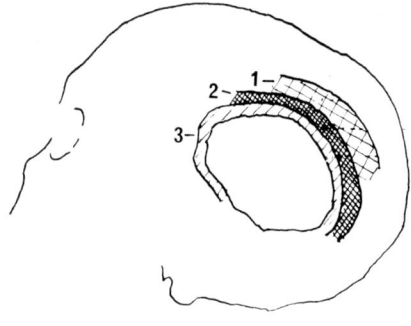

63. Diagram of a Neolithic skull, found at Nogent-les-Vierges. After Guiard (66).

1. Zone of osseous rarefaction.
2. Zone of condensation.
3. Zone of the lip.

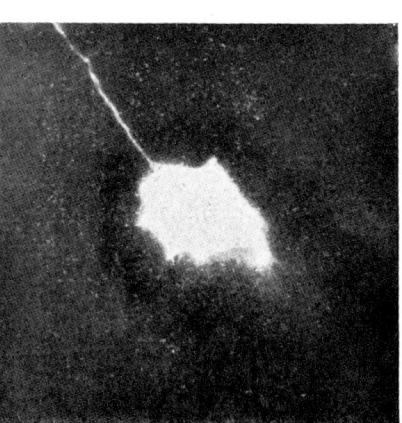

64. Radiograph of a trepanation opening. Cocarde picture. After Morel (127).

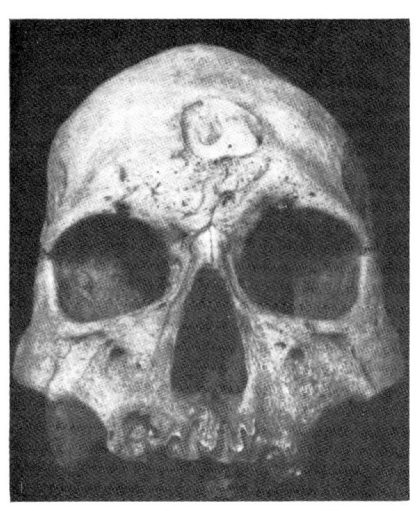
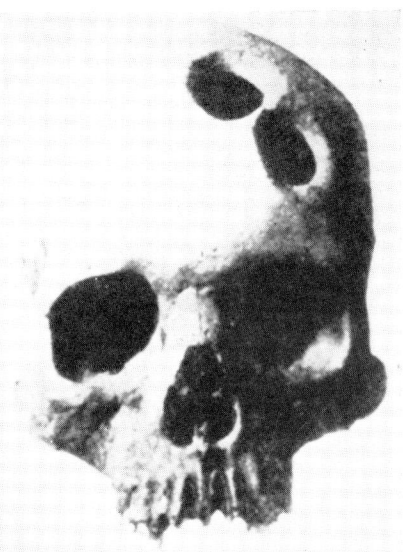

65. (left) Pseudo-trepanation. Peruvian skull with a big fistula as a result of pansinusitis. Moodie, *Paleopathology*.

66. (right) Front view of a Pre-Columbian skull, with five trepanation holes (region of Patallacta). After MacCurdy.

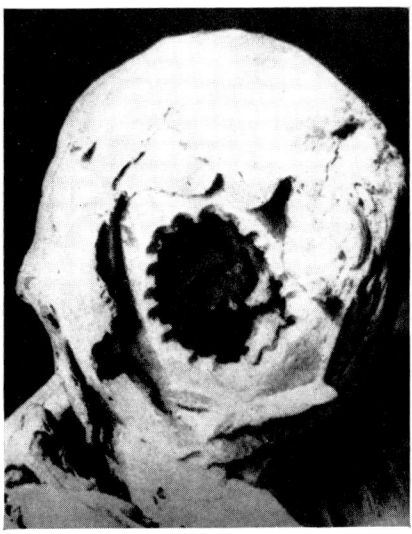
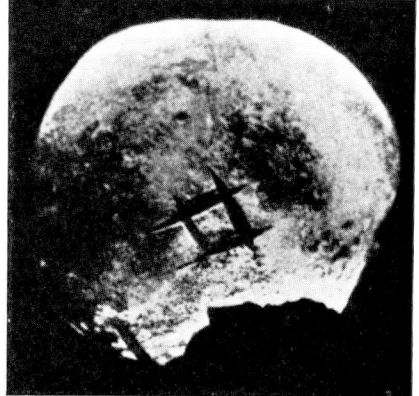

67. (left) Peruvian mummy with trepanation hole, made according to the technique suggested by Lucas-Championnière. From *Les Momies*.

68. (right) Trepanations made by a sawing technique. Yanyo skull. After d'Harcourt (40).

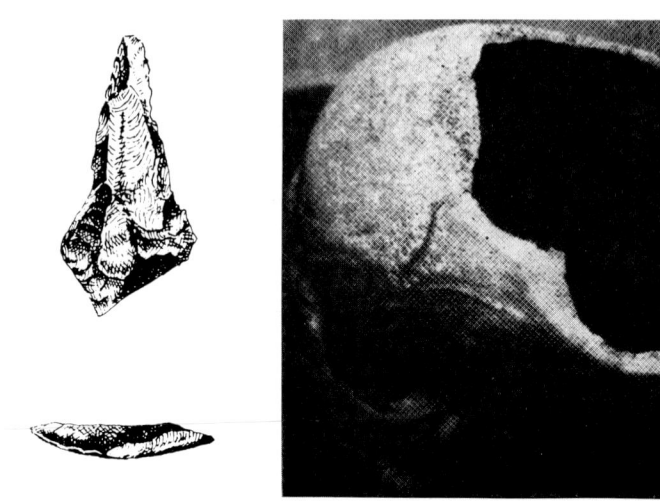

69. (left) Flint compasses. After Dr Muratet. Guiard.

70. (right) Skull with enormous trepanation. Saint Urnel. P. R. Giot.

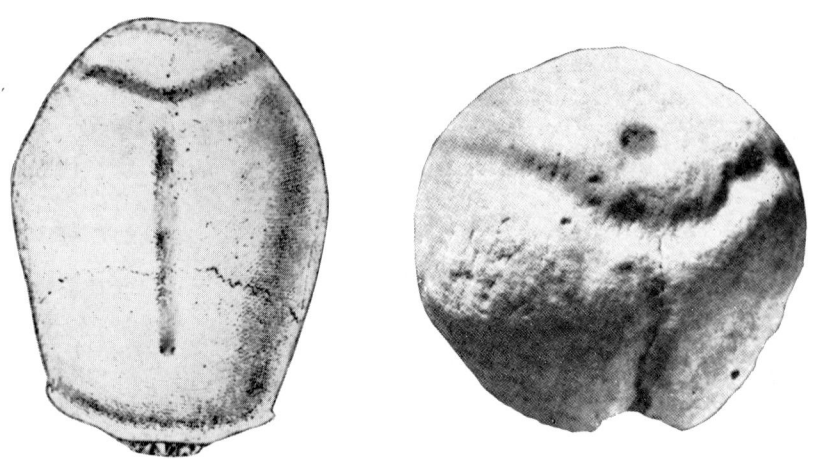

71. Skull with sincipital-T. After Dr Stéphen Chauvet (29).

The sloping lip is also found in trepanations performed by chiselling away bone fragments with some sort of flint blade. After a wide groove had been made in the bone, the bone fragments were dug out with a knife, which was held obliquely. In this way there was less danger of injuring the dura mater. A similar technique is one in which the bone is dug out, splinter by splinter, with a flint blade.

Finally there is the technique in which a square piece of bone is sawn out. Such a trepanation, which had healed, was found by Périaire and Terrier in a skull from Lizières. A few similar ones have been found in Peru. H. Müller repeated Broca's experiment on human skulls and through scraping made openings in thirty-one minutes, thirty, and sixty minutes. He also completed a 'rondelle' by sawing in seventy-five and 115 minutes. The latter method was mainly used for post-mortem trepanations (Pl. 68).

Dr Baudouin put forward the hypothesis that it was performed using a flint compass (Pl. 69). This object is shaped in a crescent. One point was used as the support, and the other cut while rotating around this. Rondelles with double perforation were made through performing the same technique twice, each time with differently placed point of support. This hypothesis is not very sound, as there are rondelles with eccentric perforation. Besides, only one similar instrument has been found, in Magdalenian layers, though it was used to cut circles. Further, Dr Baudouin considers that trepanations with a lip-shaped edge were performed by the Neolithics with the sole purpose of obtaining bone powder. He also is convinced that this form of trepanation was only performed on living people.

In MacCurdy's description of the Peruvian skull with five trepanation holes attention is drawn to a sort of incomplete trepanation which involved only the tabula externa. He does not discuss the matter in detail. Morel (126) describes a fragment of a left parietal bone with a regular circular cut, situated very near to the sagittal suture. It has a diameter of 13 cm. and a depth of about 6 mm. Only the tabula externa and the diploë have undergone a loss of bone, the tabula interna showed no lesions. The wall of the hollow was covered by a thin scar tissue layer. The author wonders whether this is an incomplete trepanation in the sense that Broca speaks of 'les trépanations des esprits forts'. Morel nevertheless considers that it could be a lesion caused by a bruising object, where the bone sequester has been expelled.

Rondelles were discovered for the first time in 1867 by Prunières, and described in 1874. Sometimes they are polished, or perforated, and often the edge shows part of a healed trepanation hole. In a few cases they have been found in healed trepanned skulls, where we get the impression that this was meant to be some sort of restitution to the survivor. Others regard them as objects which were used to cover trepanation holes, as some primitives still do with the shell of a coconut (in the Loyalty Islands).

Dr Baudouin regards them as a symbol of a stellar-solar cult and supports his fantastic hypothesis with examples from Tibet, where a necklace has been preserved, consisting of 106 rondelles from living priests. He even regards them as the forerunners of the rosary.

A rondelle is a fragment of bone with steep edges and open diploë. Its size is about 4 or 5 cm. The oldest known example is made from the parietal bone of a child, and has been described by Lagotala; it was found in a Magdalenian settlement at Veyrier near Geneva. The fragment measures 4×6 cm. and shows no signs of healing. Sometimes rondelles are irregular, triangular, trapezium-shaped, or rectangular. They were obtained through sawing, and in a few cases simply through breaking a fragment of bone. In some cases one or two holes show that the rondelle has been fastened to a string; in others there are no holes but a groove may replace them.

In Belgium we have a Neolithic skull from Hastière, with a post-mortem trepanation. The skull was that of a man of about forty to fifty. Houzé describes the lesion as follows:

In the obelionic region, two and a half centimetres from the lambda, there is a uniform loss of tissue in the form of a shield and measuring 20×25 mm. On the sagittal side, a fragment of bone has not been separated and bears traces of saw-cuts.

One perforated rondelle has been found in the cremation burial place of Overpelt (Limburg, Belgium) (136).

Another cranial operation, related to a rondelle is the enlarging of the hole in the back of the head by deepening the dorsal edge. The purpose of this operation is not known, though it often occurs. For instance it occurs in all skulls of Sinanthropus found in the cave of Choukoutien, of Samboang, and in the skulls of proto-Neanderthals of Steinheim and Weimar, in the skulls of a child from La Quina (Charente), and of Pech de l'Aze in Dordogne, as well as in those from San Felice Circeo. Besides these all Sinanthropus skulls were fractured. Without wanting to study the matter in detail, I mention that according to H. Breuil (19) this operation does not necessarily point to cannibalism. Von Koenigswald (179) does not agree and regards it as a certain proof of cannibalism.

What was the purpose of post-mortem trepanations and even of trepanations performed on living people? One reason for post-mortem trepanation could be a practical one. The hole could have been made to allow the skull to be hung up, as is the custom among the Dyaks of Borneo, where a cranial cult is still observed. Others think the purpose of the operation was to transform the skull into a drinking vessel, possibly for ritual use. 'Boire dans le crâne d'un ennemi est la volupté suprême du Barbare', writes Broca, after a text of Titus Livius (Book XXIII, chapter xxiv).

Lehmann-Nietsche imagines that trepanation was performed to allow removal of brain. Cartailhac is also of this opinion and even supposes that after extraction of the brain resin-like substances were put in the skull to preserve and protect it. Indeed such substances have been found in a skull, which is now in the museum of La Plata. No satisfactory explanation exists for resection of the hole in the back of the head performed by the Ainu. Perhaps this too was performed to allow the removal of brain tissue. The Japanese regarded this operation as a remedy against syphilis. Virchow regarded it as a wound caused by a cutting object on a living person.

Finally, some investigators regard trepanation as a surgical intervention which allowed the obtained bone fragments to be used as a remedy. They base their view on the fact that right up to the last century chemists sold 'ossa wormiana' as a remedy for epilepsy. Their triangular shape is suggestive of the irregular rondelles.

The many different possibilities to explain the purpose of trepanation have resulted in different classifications. Broca and Prunières spoke of trepanation for surgical purposes and of post-mortem trepanations. Guiard spoke of trepanation, followed by recovery or not, after Smith had drawn attention to the fact that a so-called post-mortem trepanation could easily have been performed during life, and followed by the patient's death during or soon after the operation. Le Double distinguishes between trepanations for surgical purposes ('trépanation chirurgicale') and trepanations for medical purposes ('trépanation médicale'). The first type was performed in cases of osteitis and necrosis of the cranial bones, hernias of the encephalon, and in cases of hydrocephalus. The second type in cases of epilepsy, hysteria, delirium, convulsions, and madness. Because of the long time it takes a trepanation to heal, Broca assumed that it was performed exclusively on children, especially as they suffer more from convulsions than adults. This thesis is not sound, however, as trepanned children's skulls are rare and only adults show non-healed trepanation holes. Broca based his hypothesis on the confusion which existed between epilepsy and convulsions, which is very clear in Jehan Taxil's writings at the beginning of the seventeenth century (165).

It would seem that this confusion was behind the performing of trepanation in many cases where what might have been considered as causal factors were in fact irrelevant. It is understandable that primitive people can regard a demonic spirit inside the patient as a causal factor for diseases of the mind, involving convulsions or not. Even today we still speak of someone 'possessed'. To deliver the sufferer – and thus to cure him – a trepanation seemed to be the required operation.

In Broca's time avitaminosis D was completely unknown as a pathological entity, though he did make the connection between convulsions and rachetic teeth. At the same time he rightly remarks that real epilepsy only occurs after the tenth year, and that Neolithic people could easily have ascribed all convulsion to epilepsy, just as Taxil, for instance, did in the seventeenth century. We might add that not all convulsions in children are caused by avitaminosis D. Some, called *Fieberkrämpfe* in Germany, originate from ordinary hyperthermia, whether malignant or not, depending on the aetiology. Thus the child suffering from periodic convulsions may have undergone trepanation with a good result, contrary to a sufferer from epilepsy. But as the latter affection is rarer, the overall statistics still turn out in favour of trepanation. It is striking that avitaminosis D occurred in the Neolithic period for the first time and that in the same period trepanation is also performed for the first time. An additional factor, which is quite important, is that epilepsy is often regarded as a 'holy' disease, as are all mental illnesses. Trepanation would perhaps have been considered here, especially bearing in mind what has been said about the confusion between convulsions and epilepsy.

The high percentage of healed trepanations in times when asepsis was not known is very surprising (144). We can even say that healing was normal. Of ten trepanned skulls from Baye's collection there are nine with healed trepanation edges. In Peru MacCurdy found eight cases where trepanation of fractured skulls led to immediate death (113). In eleven cases there was partial healing and in twenty-six others there was complete recovery, while in two cases the post-operation course could not be traced. MacCurdy concludes rightly that trepanation was not such a dangerous operation.

From this we can conclude that prehistoric people certainly knew a family life, and that the ill and wounded were cared for. At the same time we see that they had strong resistance against infection (Lehmann-Nietsche) and that large trepanation holes outside the cleavable zones and lesions of the dura mater did little or no harm. This is illustrated for example in a skull found in a common grave of the tumulus of Kersaint-Plabennec (Brittany), and preserved in the Museum of Penmarck: the trepanations form one large opening. which had taken away the greater part of the cranial roof. The bones involved are both parietals to the temporo-parietal suture on each side, the frontal bone to beyond the bregma, and a small part of the occipital bone. The border is formed by several bridges which show a steep edge of fresh bone, or a sloping lip, which is more or less cicatrized. There can be no doubt that here several trepanations have been performed. P. R. Giot (59) described the lesions as follows (Pl. 70):

The total length of the opening (measured along its chord) is 140 mm., its breadth is 110 mm. The first two operations were performed on the left parietal eminence where the arcs of two trephinations intersect. The diploë shows sealed-off cancellous bone and the internal table proliferative repair. Another goes from the left coronal suture, across the frontal bone, to find itself cutting again into the anterior third of the right parietal. Although open, the cancellous tissue of the diploë has begun to proliferate longitudinally in such a way that we can infer that this operation was done later than the others.

At the level of the occipital bone two arcs appear which almost intermingle: the first involving a small part of the left parietal and the occipital, the second extending as far as the posterior third of the right parietal. The bevel of the edge is reduced; scarring is advanced, with extensive irregular repair.

The remaining part of the right parietal reveals an arc with a clean cut border leading into another which is jagged. There has been an ante-mortem resection here, rapidly followed by death, or else the bony fragments became detached after the person's demise.

Condensation or rarefaction zones could be traced radiologically on the left parietal and the frontal bone but are missing on the right parietal bone. The whole seems to suggest a tumorous process of the brain with a fast course, such as sarcoma, but the facts and indications for such a causal process are missing.

Although it is agreed that trepanations on living people were performed in the Neolithic period, there is no agreement about the purpose of this operation. We can distinguish three points of view. The first consists of those who think trepanation was primarily a ritual action; Lecène even rejects all medical or hygienic grounds. De Mortillet regards it as a privilege, a sign of dignity, perhaps like the tonsure. He even sees trepanation as a forerunner of the latter. Dr Baudouin's hypothesis has been mentioned already: trepanation provided bone dust or rondelles – and is connected with a stello-solar cult. It should be pointed out that ethnology informs us that trepanation performed by some primitives today, never has a ritual significance.

The second consists of those, like Le Double, who believe they were performed for medical reasons (107). Here should be included Broca's hypothesis that trepanation was performed on children suffering from epilepsy. Le Baron agrees with this view and even supposes that when the patient survived, he was venerated. Lucas-Championnière also endorses this view, supporting it by the fact that in antiquity epilepsy was regarded as a 'holy' disease. For this reason Morel even wonders:

whilst admitting the religious character of trephinations, whether they were not intended, instead of seeking a cure, to provoke in those who were thus singled out, precisely those psychic disturbances and convulsive crises which revealed the immanence of the god.

He illustrates this with the number of cases of epilepsy following cranial wounds in the first world war, which varied between 12 per cent and 37 per cent. He realizes that there would have been more 'bad' than 'good' results, but he thinks that pious fervour, suggestions, and simulation took care of the rest. Gastaut (54) produces corresponding figures: 30–40 per cent for the first world war, and 10–15 per cent for the second world war. Walker (181) gives 25 per cent for the second world war. I think these figures alone are hardly convincing enough to justify the supposition that trepanation was an 'epileptogenic treatment' of the Neolithic period.

Prunières follows a medically more sound course of ideas: he thinks that in the Neolithic period cranial fractures occurred more frequently, and as a result also traumatic epilepsy. The removal of bone fragments after the trauma prevented this complication, and because of this trepanation was also performed to cure other convulsive affections. On the other hand Neolithic trepanned skulls with stigmata of a fracture are rarely found. Prunières himself mentions a child's skull with a depressed parietal with no surgical intervention. In Peru cranial fracture is definitely an indication for trepanation. MacCurdy found fractures in 22 per cent of the men out of a total of 130 male skulls, in 15 per cent of the women out of a total of 108 female skulls. In both sexes he found thirteen cases of trepanation after fracture, once to remove such bone tissue, thirty-one cases which showed no signs of disease or wounds, and two cases of which the cause could not be detected because of *post-mortem* destruction of bone. If MacCurdy finds trepanation a sign of impacted fractures we can assume that the figure is still higher because there must have been cases in which the trepanation hole caused all sign of the fracture to disappear.

These remarks have already brought us to the third viewpoint, which finds in certain trepanations some surgical indication. Besides the indication of impacted fractures I mention Lucas-Championnière's view which treats trepanation exclusively as decompressive intervention. This raises the question why in so many cases we find more than one trepanation hole. Or did the Neolithics not realize that the decompressive effect can be reached simply by reopening the soft parts of the first trepanation?

The skull of Lizières may have been trepanned because of osteitis. This could also have been the case for the skull from Bray-sur-Seine, described by Parrot. Broca mentions the operation on a hydro-cephalic skull. Williams found a few skulls in Peru with signs of symmetric osteoporosis, while Morena found luetic lesions on a skull from La Plata. Carrière and Reboul found several trepanned skulls with serious traumatic lesions, which were frequent in ancient Peru, as MacCurdy also mentions. Moodie ascribes many of these impacted fractures to slingstones. This traumatic factor set Wölfel off checking up on the geographical distribution of the use of this weapon and trepanation. He found a striking connection between the two factors. It can be seen also that trepanation is practised among brachycephali and not among dolichocephali, which we find in England, Portugal, and Spain during the Neolithic period. An exception are the dolichocephali from Melanesia. I have mentioned in the introduction the trepanned skull from the Mesolithic period in Spain. The particular phase of this period in that region is called the Asturian. The complete skeleton was found under a tumulus at Colombres, near the well-known cave of Pindal (25). It is extra-ordinary that the third molar, on the left of the lower jaw, showed dental caries, and that the skull is distinctly dolichocephalic. The sex could not be determined with certainty. The trepanation was performed just before death or *post mortem*. The opening is oval, the size of a penny. It is situated on the left temporo-parietal zone, and a small part of the sphenoid was also involved in the opening. There were no traces of healing at all.

Stéphen-Chauvet (29) draws attention to the fact that in most cases of impacted fracture of the cranium, caused by blows from polished axes, the lesion is situated on the left side of the skull, usually on the parietal and temporal, as a result of the attacker being right-handed. He explains certain small trepanation holes as a means to avoid causing cracks in skulls with impacted fractures. The trepanation hole was made with a flint knife, provided with a triangular point.

Trepanation survives the Neolithic period but becomes rarer. We continue to find it among Gauls, Franks and Merovingians. Cases have been described by Schmidt (156) in men and women from graves in Central Germany from the second half of the fourth and beginning of the fifth century. Nothing is known about possible anaesthesia, but we can assume that it was superfluous in most cases because the patient was likely to be in a coma. When this was not the case we might take into account that primitives are perhaps less sensitive to pain. Finally it is quite possible that an alcoholic drink or narcotic decoction was used, as is still the case with some primitives.

I consider that the ritual factor as an explanation of trepanation should be rejected outright. Giot (60) emphasizes a peculiarity of a common grave near St Urnel-en-Plomeur where trepanned skulls were found: there were almost no grave-goods, while they occur plentifully in tumuli of the same period situated no farther off than a few hundred metres, such as those of Kervlitrez, Roz an Tremen, and Tronoan. Thus there are no signs of any particular veneration as we should expect for possibly ritual trepanations. It could almost be called a common grave for special cases: not massive epidemics, caused by one or the other pestilential disease, but victims rather of small epidemics – the young age of the victims points to this – or perhaps the mentally diseased who had been trepanned. The fact that the skeletons lie near to one another and that already existing graves have not been respected also points in favour of this view.

SINCIPITAL-T. BREGMA-NERVEN

Associated with trepanation is another peculiar lesion of the skull, described by Manouvrier (115) in 1895 as sincipital-T in about six female Neolithic skeletons which had been buried in the dolmen of Epône at Nantes (Seine-et-Loire). A similar case is mentioned in America by Roy Moodie (129). A few years ago Weiss (182) took up again the study of this lesion. He mentions a total of twenty-four skulls.

The lesion consists of two grooves which cross each other: the front-to-back groove starts from the frontal bone and extends to the lambda along the sagittal suture; the transverse from the left parietal to the right. The sagittal groove is always present, but the transverse groove is missing in some cases. Sometimes also the transverse groove is like a broken line so that a Y-figure is formed (Pl. 71).

The groove is usually restricted to the tabula externa, sometimes to the diploë. The edges are caused by reactional thickening of the exocranium and are 2 to 3 mm. deep and 1 to 2 cm. wide.

The groove from front to back is narrow and suggests an incision, not a cauterization. The transverse one is not very noticeable in the middle and becomes deeper towards the sides of the skull and ends on both sides in a groove of about a finger wide which sometimes perforates the tabula interna in an irregular perforation of 3 to 4 mm wide. It has cutting edges. The groove is presumed to have been caused by cauterization. Sincipital-T is rare compared to trepanation. Cicatricial exostosis, following cauterization, has in the past often been confused with osteomyelitis. Cauterization is found only on skulls of females and children, never men, while trepanation is almost exclusively performed on men. The fact that the operation was usually successful is proved by the fact that of twenty-four skulls examined by Weiss, only one did not show any healing reaction. Baudouin (11) mentions one case of trepanation and 'rainure sagittale' on the same skull of a man from the ossuary of Vaudancourt.

The reason for performing the sincipital-T is not known. Frederick Grön regards it as a punishment. Others associate it with magic, ritual, and initiation practices. Since sincipital-T occurs only in women it has been thought that this operation was a substitute for trepanation or that it was performed in less serious cases. Sudhof, Le Double and Weiss regard it as an operation with great revulsive power, which had been passed down by way of Hippocrates and Galen, as treatment against headache, mania, melancholia, epilepsy, convulsions, and delirium. Similar practices were in use in the Canary Islands, where the Guanches made deep gashes in the skin of the affected parts with their stone knives and afterwards cauterized with roots of malaccarite, dipped in boiling goats' fat. Alexandrian surgeons made deep lacerations on the forehead against certain affections of the eye, while the Arabs cauterized the cranial skin as treatment for epilepsy, melancholia and some other mental diseases. Finally, Krogman (100) regards sincipital-T as no more than a form of tattoo.

# CHAPTER 27

# Therapy

Since we can trace back the presence of disease to a time well before man's appearance on earth, we might wonder whether mankind reacted immediately against it – in other words, whether medicine came into being at the same time as man's becoming ill. I have already mentioned this problem in the introduction. We should now consider how far the notion 'medical deed' stretches. Does a doctor commit a 'medical deed' when he cuts the umbilical cord at birth? If so, medicine has existed certainly from the first man on earth, and it was – and still is – practised by animals. However, we want to specify medicine as the medical or paramedical help given to a diseased man by his fellow men. There is no absolute certainty about this being so in the Palaeolithic period. H. Kuhn (102) mentions the engraving on a bone plaque found at Limeuil (Dordogne), in which he thinks he can recognize a sick man, lying on the ground, with a medicine man bending over him, exorcizing the illness.

We can guess at the existence of medicine, because we know that burial and cremation existed; is not burial a sort of extended help, which commences with sickness? Further, the burial rite develops alongside the development of the race. Burial is the mental link between the living and the dead.

Even the Pekin man seems to have had a burial ritual. Only the skull has been found, amidst many more fragile fossils, so we might assume that burial of the head alone took place. This form of burial occurs among the Papuans today (96, 177). Others have suggested cannibalism, as it must have existed among later prehistoric races, because most skulls have been found smashed to pieces and with an enlarged occipital hole (36). The most flagrant examples are those from the cave of Ofnet (Pl. 72) (10). This cave is situated near Nördlingen in Bavaria. In 1908 R. R. Schmidt made important discoveries there: the so-called 'skull nests'. In two pits a number of human crania were found, which, doubtless for religious reasons, had been buried neatly packed. The pits were richly covered in ochre powder. The skulls all faced westward. The larger pit had a diameter of 76 cm. and contained twenty-seven skulls; the smaller one contained only six. The skulls showed lesions caused by axe blows, and the cervical vertebrae betrayed that the heads had been severed by flint knives. Of the skulls, eight were dolichocranial, eight mesocranial and eight brachycranial. They represented nine women, twenty children and only four men. The women and children all had gifts in the form of perforated teeth of deer and perforated shells, which probably had decorated their headgear. The men had received flint tools. The absence of a proportionate number of male skulls suggests the murdering of a small tribe when men were out hunting. Whether cannibalism was involved is impossible to say. Yet the killing was followed by a special burial of the heads.

The skulls of Ofnet have always been regarded as dating from the Mesolithic period. I think that in this case anthropology and even more palaeopathology should be of help to the archaeologist in dating. First, the different forms of cranium: pronounced cross-breeding exists here. Secondly there are the holes in the skull: what weapon could have caused them? The hole is lens shaped, and is certainly too big for an antler-axe of the Lyngby type. In any case the wound has a straight side. Stone axes, in the real sense of the word, did not exist in the Palaeo- and Mesolithic periods: they occur only in the Neolithic period. My opinion is then that these are a Neolithic populational group, buried in a Palaeolithic layer.

In Predmost, Absolon found a skeleton placed on the fire hearth. The bones had carvings on them, and the head was missing. Dupont (44, 46) assumes cannibalism for the Neolithic race of Furfooz and justifies his assertions by pointing to a few small scratches found on certain bones, belonging to eight individuals, as well as to impact lesions in the ilia. As has already been mentioned, I think these lesions were caused accidentally after death by later burials in the same cave and by rodents and beasts of prey and have nothing to do with cannibalism. The occipital hole of the skulls showed no trace of enlargement. Among recent primitives a so-called cranial cult still exists (179).

In Europe cannibalism lasted for a long time. The historian Strabo tells us how the inhabitants of Ireland held it in high honour to eat the body of their deceased parents. The primitive imagines that through this deed he will inherit the qualities of the deceased.

Neanderthal burials have a pit, in which the head lies on a stone, sometimes facing west. There are no actual gifts, except sometimes a hand-axe.

Cro-Magnon man edged his grave with stones. The death bed is covered with ochre, probably regarded as a replacement of blood for life hereafter: indeed the dead are pale and appear bloodless. The feet are usually bound together in a crouching attitude: this represents the fear of the deceased returning. This posture, called the foetal posture, has been interpreted by Senet (158) as the desire to give the deceased an ideal resting position or as a preparation for a rebirth. Amulets, ornaments, tools and food are provided in the grave. Smashed skulls point to gerontocide or cannibalism. Among the Cantabric inhabitants cremation must certainly have existed, for material is so scarce (24) that only one skull of the Cro-Magnon type has been found there, in the cave of Santian (1). I think there is room for serious doubt about this being a prehistoric skull – although it is described as having Neanderthal characteristics, which for us are simply not present at all. Only a few human remains of no particular significance have been found here and there in the cave.

From these indications we might conclude that the idea of a future life was more material than spiritual, in other words that prehistoric man did not have any notion of the soul: that he did not think abstractly. That his high level of art should only be compared to the dexterity of some children today. I cannot really agree with this view. Not everyone agrees on the artistic value of prehistoric pictures: I have already emphasized this in considering such art.

To prove that prehistoric men could not think abstractly, some investigators have based their view on the fact that these people did not even see the connection between the drying out process of clay and the baking of pottery. Still, Sigerist (161) regards the 'Venus' figurines as a representation of the 'Great Mother'. An engraving of a reindeer,* standing over a pregnant woman (Pl. 73) may point to the idea of magical induction of the strength of the animal into the mother to be. Manduit (120) relates how this action is regarded as predicting an easy delivery in Baltic regions, although he too, agrees with magical induction. Mrs Della Santa (154) mentions that others have given this figure a totemic explanation: indeed the hair-growth of the pregnant woman is very clearly noticeable. The idea may be that the reindeer caused pregnancy –

---

* There is some doubt in the minds of many archaeologists that the animal is a reindeer or that there is any correlation between the animal and the woman.

that is indirectly, without any thought of bestiality – and thus becomes the totem ancestor of the unborn child. She bases this view on analogous facts among Australian tribes, where the intervention of the moon and raven cause pregnancy, and where the tree lizard begets sons. Here another question poses itself: does the primitive really understand the process of procreation? In other words, does he not perhaps regard as necessary the mental intervention of the animal which will become the totem? Although the primitive, according to Nieuwenhuys, realizes the connection between mating and nesting, this may not seem true of the human, where all sexual intercourse does not necessarily result in pregnancy. The essential difference between humans and animals lies in the fact that the latter have an oestrus period. Man sees the connection between sexual intercourse and pregnancy but he cannot give a causal explanation of the embryological process and will call on an 'intermediate host' or totem. This ignorance in fact only disappears with the discovery of the De Graefian follicle, and it makes its influence felt right up to historic times in Greece, where it was still assumed that the woman only fed the child, which was brought into her womb by the man, without her taking any real part in the initial begetting of the child. After the capturing of an enemy town the women were regarded on equal standing to the victors, without influencing the purity of their race, so that they were usually spared. This was no longer so when it was assumed that she took active part in the forming of the fruit herself.

In order that the reindeer could be a totem, there should be an alimentary taboo, or the animal should occur a lot in one particular place. This does not seem to have been the case. At the same time it should be linked to a 'clan', which was possibly so, but is difficult to prove. A relevant example is the representation of Lascaux where a human is pictured with a bird's head, together with a pole on which a bird is seated. According to Seuntjens this is a totem figure (159). His theory is that the male figure is not lying on the ground, when he would be regarded as dead or wounded: exactly the opposite – the man is standing in front of the bison and exorcizing it: everything depends on the angle at which the photo is taken because the rounded shape of the rock wall can cause an optical illusion.

The human figure is often linked to the animal one, sometimes almost invisibly. The impressive ceiling of paintings at Altamira consists, besides animal figures, of eight engravings of anthropomorphic figures, two of which have bird's heads, and six undefined animal's heads. All figures are striking because of their pronounced phallic characteristics (27). This is also to be found in the pictures of Lascaux.

I have already mentioned how far Manduit (120) takes his interpretation of the 'Venus' figurines: a fertility symbol, not for man, but for the hunted animal. Luquet (112) also takes this view. Human eroticism watches over man's multiplication and seeks to make its influence felt in the same way on the fertility of animals. Manduit reaches this conclusion by applying a line of Nietzsche's: 'We live on the remains of the feelings of our ancestors.' Mental development is not very different from bodily development. As a result, psychology of the child could be compared to the psychology of the primitive man. The latter did not reach full mental maturity. In *La Génèse des Espèces Animales* Lucien Cuénot writes in a similar vein: 'In relation to the Anthropoids, man exhibits a slowing of growth and genital retardation.' The mind of modern man still shows traces of the different stages: the first developed and deepest situated parts of the brain are closely connected to instinctive actions. Only later does the more extended grey cortex develop, the seat of conscious psychic life. Every human goes through all forms of bodily life from the unicellular onwards through metazoa, fishes, amphibians, reptiles, birds, and lower mammals during embryonic life, but the mind of man goes through all the stages during the first years of life: to understand the psychology of prehistoric man we should study the child's psychology.

In the first three years the babbling child will try, without being conscious of the social factor, to assemble his impression of the outside world in the function of his personality, which appears gradually from the unconscious. This condition is sometimes called the 'prelogical condition'.

Around the third year the child forms his own fairy world. His powers of observation and concentration are too small to allow him to make a distinction between the outside world and his own personality. His notion of conscious being is incomplete and he lives in an animated world (in which some primitives still live) where beings, things, and even natural phenomena are personified or take on a more or less benevolent appearance, which he continually tries to get within his grasp. This dream world disappears as well. The child's sensations take the form of a personal memory. He realizes the difference between observation and imagination and confuses it less and less with the dream-world, which his imagination presses upon him at the expense of reality. He is the conscious observer of his own sensations.

These few observations about the primitive and his psychology will have to suffice. I only wish to stress the personification, natural to the primitive, which he extends to disease. He searches for a concrete explanation for disease and in doing so always returns to the cause. To him it seems possible that disease is a punishment from an angry god because of some unforgivable offence. If so, there is a danger that any attention given to the patient may transfer the punishment to other members of the tribe: the only possible way out is to avoid the diseased, as happens in case of an epidemic disease, for which the primitive knows the obvious explanation. The diseased is dead socially before bodily. In other cases the patient will be surrounded with special cares, if the disease was contracted involuntarily as a result of one or other devilish dealings.

The economy is also a decisive factor in fixing an attitude towards the sufferer. The treatment may fail. The diseased has become a useless part of the tribe, a burden; his elimination is desirable. In New Caledonia the hopelessly sick are left to die or are killed with a blow on the head. The Zaparos of Ecuador strangle their hopeless cases. Cannibals, such as the Bobos of the Western part of Sudan, kill and eat their diseased before they have lost too much weight. These customs seem to contradict the accepted notion that we are dealing with an agricultural people with little food-shortage. Tribes of hunters and fishermen usually treat their diseased well. They only use these extreme measures in time of dire food shortage or migration.

A discussion of the strictly magical intervention to cure disease does not belong in this account: recapturing the soul, expelling evil spirits, annulling magic through counter-magic and so on. We have to mention them though, to avoid a false picture about some actions of primitives: if a primitive does not urinate in a river this is not a hygienic measure: he only wants to avoid offending the river spirit. If he buries his faeces it is to avoid this 'strictly personal' object being used against him.

I have already mentioned the close connection between religion and medicine, in particular in case of serious affections. Small, daily affections are not important and need not be explained. The patient himself treats them. There exists a close relation between his daily life, turned in on itself, and his inborn instincts. His actions are instinctive. He finds out what he can eat when he is healthy and what he should eat when he is ill. There is no strict dividing line between food and medicine. Certain food taken in great quantities can act as purgative, cholagogue, or as healing factor in case of vitamin deficiency.

It is clear that medical deeds can come about instinctively: some-one who bruises himself involuntarily rubs the painful spot. Worked out to a system this becomes massage. The warmth of a fire will soothe many pains. Often in the near vicinity of prehistoric graves remains of fires are found, as if it was the idea to warm the dead in their graves. Sick animals eat certain varieties of grass, and wounded animals go and sit in the mud. They also use mud to get rid of parasites. This action is the basis of hygiene. We can further mention the licking of wounds (we know the bactericidal action of saliva), the removal of thorns, and pressing on haemorrhages. The letting of blood seems to be a very primitive medical deed, which was learned from experience: a bleeding nose or menstrua-tion bring relief in feverish affections.

The sexual urge asks for well formed, healthy partners. This encourages young people to keep in good health through taking care of the body, even if the only motive is to be attractive to the other sex.

Even though these considerations have taken us a bit farther from the starting point, they give us no absolute certainty. An indication of the existence of medicine in the Palaeolithic period does not exist. It has been pointed out how dangerous it is to compare Palaeolithic men to primitives today. The latter are often decadent, having come down from a higher culture to a lower degree of civilization: their culture today usually does not correspond any more to their original one. The contrary is true of the Palaeolithic period. Another point is that even the primitives today have very different conceptions. In the Fiji Islands, the New Hebrides, and in some parts of Africa mentally ill people are buried alive, to make certain of destroying their dangerous spirit with the body, while some American Indians treat their mentally ill well and with respect, because they regard them as bearers of transcendental power. More important still is the relation between religion and medicine or between magic and medicine. Perhaps both factors should be mentioned together, remembering Rivers' definition: magic is a group of actions, ac-cording to which man follows a rite, the effectiveness of which depends on its own power, which is inherent or ascribed to certain objects and processes that have been used in the rite. Religion is an activity whose effectiveness depends on the will of a higher power, whose intervention is asked for through rites, entreaties, or re-conciliation. Religion implies a belief in a power in the universe, greater than man's power.

Having mentioned the possible psychology of prehistoric man, I now turn to his possible religion. H. Kuhn (102) thinks three

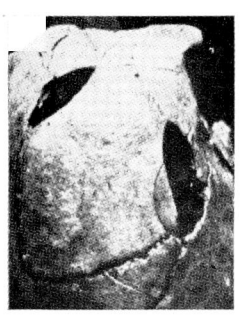 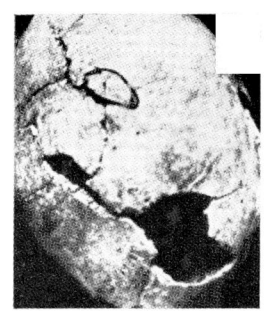

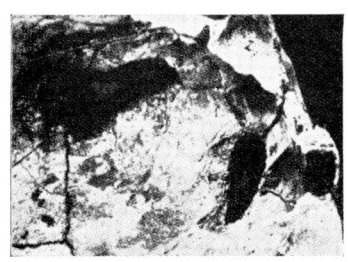 

72. Ofnet skull No. 21. Man. Top row: lesions viewed from the outside. Bottom row: lesions viewed from the inside.

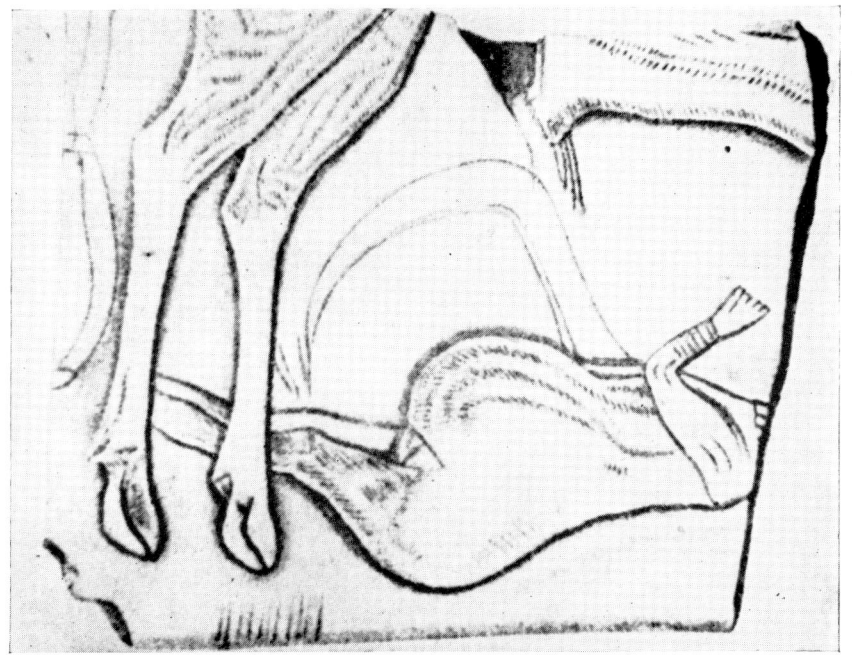

73. Reindeer standing over a pregnant woman. From Van den Broeck. *De Dageraad der Mensheid*.

74. Dr Gaussen showing his excavation of a Magdalenian hut with flooring. He is indicating the entrance and workyard on the southern side.

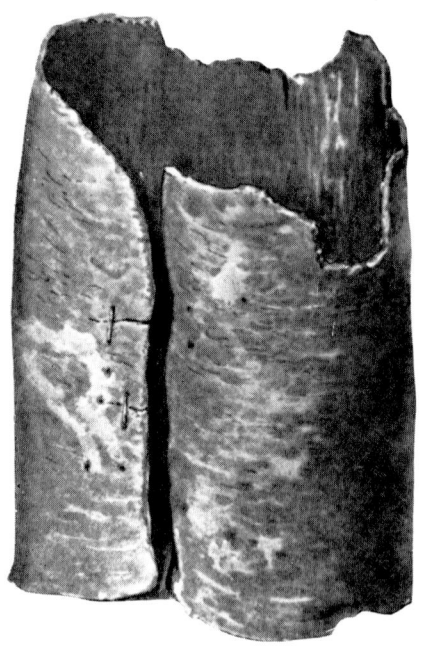

75. Corset of bark (orthopaedic?) without doubt used by North American Indians for lesions of the spine. After Freeman. From Moodie, *Paleopathology*.

factors should be considered which together determine the appearance of religion: the idea of eternity, Cult, and of the representation of God.

The idea of eternity he finds in the manner of burying the dead. He also points out that there is not the least notion yet of the division of body and soul: the deceased lives on as man, although in another form and in another place. And in order that he will not return in his original shape, his legs are sometimes tied together.

The Cult shows itself in engravings and rock paintings. The art of the Glacial period is entirely based on religious consciousness and corresponds to religious art in the middle ages. Dance and music link up with this, as is clear from finds of bone flutes and primitive wooden hurdy-gurdies which make a droning noise when they are turned. The *bâton de commandement* is also perhaps connected. It is usually thought that they were used to straighten arrows because of the hole in them, although normally no traces of usage have been found on this hole. Authors like Goury (63) however do claim to have found such traces. This *bâton*, from the Magdalenian period, is usually decorated with representations in connection with birth.

The idea of a Divinity shows itself in offerings, not those which were made to the dead, but offerings similar to those which we have described in the bear cult. The offering obviously springs from the idea of a God: it is at the centre of magic. When the offering is left out, another stage is reached: when the Bible mentions that God did not respect the offering of the fruits of the field from Cain, a new era starts, that of cultivation of the field. The peasant is not a magician, he is no longer the man who coerces, but the man who implores. The world of Adam and Eve, the world of the Ice Age man which has gone before, has come to an end. The offering is the oldest cult and no religious action reaches back as far as the offering. It is given to a God who brings fertility in animals, which is necessary for man's existence. It expresses itself also in the 'Venus' figurines, in the mother who gives birth to the child.

'The compelling action of man on nature is the origin of magic', says Maxwell. And when he says further 'les magies primitives que nous connaissons se confondent avec les religions', our attention is drawn once more to the relation between magic, religion and medicine. The religious interpretation of disease is the most important factor in establishing the attitude of the community towards the diseased. And a primitive is, just like a child, easily influenced. He tries to find an explanation for the disease when this is serious and not an everyday complaint. Why does a certain person

147

L

become ill and not his neighbour? For him it is no coincidence. Coincidence does not exist for the primitive, for no individual with a belief. Schweitzer once said that coincidence is the pseudonym of Our Lord, which he uses when he does not want to take the responsibility.

In case of disease there is always a punishment of God behind it, or the vengeance of an enemy who through magical intervention has caused the loss of the soul or the entry of a demon, which brings disease into the organism. We shall leave the several causes for disease supposed by the primitive, while stressing the importance of their magical or religious interpretation, mentioned in the introduction as having a prognostic rather than a therapeutic value. Our own history proves the point: did not the interpretation that disease was a punishment from God and therefore should not be treated because of its purifying action, halt the development of medicine in our regions? This is in sharp contrast with medieval Arabic medicine, which flowered because of the prophet's words: 'God has not created disease, without having created a remedy as well'.

These somewhat drawn-out considerations might seem to be an attempt to draw a veil over our lack of knowledge of the life of prehistoric men. In the field of medicine we can, through comparison, assume that it is possible that Palaeolithic men, being hunters and fishermen, did treat their sick. But certain wrong notions about their social life still seem to linger on: first, we must not assume that Palaeolithic people were continuously wandering hunters. The caves of Altamira and Lascaux were not painted with so much care for a stay of only a few days or even weeks. The enforced living in caves was not the cause of a common tribal life: there was already a family life. Indeed, besides caves the Palaeolithics also lived in huts. The excavations by my colleague Gaussen in Neuvic-sur-Isle (Dordogne) brought to light plans of huts, the utensils from which date from the Magdalenian (Pl. 74). These huts even had a stone floor. Although they are too small to have housed a complete tribe, it is difficult to imagine that they were only used by single individuals. Family life presupposes a special care of the members towards one another: the mother's care of the child and husband; the responsibilities of the latter in providing food and obtaining the necessary materials for clothing and weapons. The disappearance of a member of the family might cause serious difficulties in the family's division of work. One of most important factors is disease or injury, which can lead to death. It seems impossible that this was not the basis for rudimentary medical help, even though we do not know this.

The transition to the Neolithic period, of hunter to farmer, must have been of vital importance. We can imagine that better economic conditions must have contributed to a better collective health, which too had a vital negative aspect. Civilization leads to preservation of the weak individual. Natural selection in the Palaeolithic period, when only the strong survived, must have had a heavy hand. Civilization on the one hand and softening of the individual on the other must have given a most important impetus to medicine in these times.

Oswald Heer (68) has studied the remains of plants in lake dwellings, and could identify 115 different botanical finds. Besides fungi and moss, which were probably used to light the fire, he found several sorts of grain, as well as vegetables and fruit, which can be regarded as ingredients of daily food. At the same time eight plants which can be used for medical purposes occurred: *Sambucus nigra* L. (elder), *Sambucus ebulus* L. (herb elder), *Papaver somniferum* var. *antiquum* L. (opium poppy), *Tilia grandifolia* (linden blossom), *Pinus silvestris* L. (Scots Pine) *Juniperus communis* L. (juniper berry), *Viscum album* L. (mistletoe), *Rhamnus frangula* L. (buckthorn). Yet plants with a specific action, such as digitalis, do not occur at all. Bubbling geysers and the peculiar taste of mineral sources must certainly have attracted the primitive's attention. Near the springs of St Moritz wooden pipelines have been found, which point to its use. The lake-dwellers would not have needed this water for daily use.

The sensible Neolithic ventured upon surgical intervention: removing arrowheads must have been an almost daily routine: large cranial fractures successfully healed. We have already discussed the possible existence of an orthopaedic knowledge. The oldest splint, however, is one found in an Egyptian grave from the Vth dynasty. Though splinting is a general medical treatment which might be discovered instinctively, this does not mean that reduction was achieved. Primitive North American Indians also made orthopaedic corsets from bark (Freeman) (124) (Pl. 75).

In view of the importance of trepanation I have already devoted a separate chapter to this operation.

Surgical instruments have been found in La Tène graves (late Iron Age) which probably point to graves of medical men of the time. They include strings, retractors, and amputation saws (191).

# Conclusion

We can trace back disease to a time long before man appeared on earth, to the appearance of the first vertebrates. We have gone the full circle: disease has always shown itself in the same way. We find the identity of today's diseases in the anatomo-pathological picture of the discovered palaeopathological pieces. Reactions to disease have remained unchanged since the time the vertebrates appeared. This is completely contradictory to MacCallum's opinion, that spontaneous defence reactions of man have developed and improved over the centuries. This is not so: man has, perhaps, managed to protect himself against disease through external means, more than through internal ones or through immunity. The latter certainly existed in earlier times and may even have been more efficient then: the more primitive a being, the more efficient its defence and recovery mechanisms seem to be. From only a part of an animal of the earlier sort, a worm or a jellyfish, a new whole can grow. Some lost organs of an amphibian and even a reptile can be regenerated. We also find that some primitives today are immune to certain infectious diseases.

For highly civilized man the time of natural selection is past. The weaker are kept alive by modern therapeutic means, while sturdy youth is decimated by war. But modern medicine has another danger: the use, or rather abuse, of radioactivity, for therapeutic as well as diagnostic purposes. More and more we realize that the human body can stand only a certain dose of X-rays, without damaging health and procreation of the individual. Rays can cause a change in the genes of the egg- and sperm-cells. Whether these mutations will be advantageous or disadvantageous for posterity is to be seen.

If the supposed promiscuity of prehistoric man, with all its regressive risks, was a homologous factor then the tens of thousands of years of prehistoric life suffered very little damage from it.

# Glossary

ACHONDROPLASIA A form of dwarfism in which the limbs are greatly shortened in comparison to the usually normal trunk.

ACROMEGALY A disease of the pituary gland which leads to excessive growth of face, hands, and feet.

ACROMION The triangular process of the shoulder-blade to which the collar-bone is articulated.

ACTINOMYCOSIS A fungal infection.

ALVEOLAR The tooth-socket. Also used in connection with the air-sacks of the lung.

ANASTOMOSIS Connection between lymph or blood vessels, or between nerves.

ANEURISM Swelling in an artery.

ANGIOMA A tumour formed of blood-vessels.

ANKYLOSIS Fusion of the bones forming a joint.

ANTHRACOSIS Miner's lung.

APONEUROSIS Spreading of the tendons.

APOPHYSIS ODONTOIDES Dens epistrophei. A toothlike process on the 2nd cervical vertebra.

ARTERIA MENINGICA The artery of the meninges.

APICAL Relating to the roots of the teeth.

ARTERIOSCLEROSIS Thickening and hardening of the walls of arteries.

ARTERITIS Inflammation of the arteries.

ATLAS First cervical vertebra.

ATROPHY Wasting away of the body or part of it.

AXIS Epistropherus, 2nd cervical vertebra.

BESNIER-BOECK'S DISEASE Infection related to T.B.

BILHARZIA Schistosomiasis, a parasite disease caused by worms which live in the blood vessels.

BOUILLAUD'S DISEASE Arthritic heart infection.

BRACHYCEPHALIC Short-skulled (ethnology).

BUBO Enlargement of the lymph glands in plague.

CALCANEUS Heel bone.

CALLUS Excess bone produced in repair of fractures.

CHLOROMA Malignant tumour. Shows a green colour when dissected.

CHOLESTEATOMA Benign tumour between the meninges and the cortex of the brain.

CHONDROMA Cartilage tumour.

CONDYLES Bony parts of the joints.

CONVULSIONS Involuntary spasms affecting the body.

CORACOID Process of the shoulder blade.

CORONALIS (SUTURA) Transverse suture.

CORPUS CAVERNOSUM Swelling organs in the penis.

CRETIN A form of dwarfism brought on through thyroid insufficiency.

CUSHING'S SYNDROME An endocrine disease associated with obesity and excessive growth of hair.

DERMOID CYST A benign cyst, where the remains of the ectoblast, such as hair, are still present.

DIAPHYSIS The shaft of the long bones.

DIPLOË Sponge-like tissue of the skull situated between the tabula interna and the tabula externa.

DISTAL The part of the limb farthest from the body.

DOLICHOCEPHALIC Long-headed (ethnology).

DURA MATER The hard layer of the meninges.

EMBOLISM (AIR) A blocking of the brain or coronary arteries by an air bubble caused through opening a large blood-vessel, or a sudden change in air pressure.

ENCEPHALITIS Inflammation of the brain.

ENDOCRINE Relating to the internal secretion glands.

ENDOSTEUM Connective tissue which covers the marrow cavity of the long bones.

EPICONDYLUS Osseous parts above the condyles.

EPIDIDYMIS An oval body situated on the testis.

EPISTROPHEUS 2nd cervical vertebra.

EWING'S TUMOUR Soft malignant bone tumour.

EXOSTOSIS Abnormal, non-malignant bone growth.

FEMUR Thigh bone.

FIBULA The lateral bone of the lower leg.

FIBROMA Benign tumour of the connective tissue.

FISTULA Congenital or contracted pipe-line canal between an organ and the skin, often the result of internal suppuration.

FRAMBOESIA A tropical skin infection, similar to syphilis.

FRONTAL BONE The bone of the forehead.

GANGRENE Necrosis of the tissue due to insufficient blood.

GINGIVITIS Inflammation of the gums.

GLIOMA A swelling of the nerves.

GUMMA Ulceration of the skin (3rd stage of syphilis).

GYNECOMASTIA The formation of breasts on men.

HAEMATURIA Bloody urine.

HEMIPLEGIA Paralysis of one side of the body.

HUMERUS The bone of the upper arm.

HYDATID CYST A parasitic cyst.

HYDROCEPHALUS Dropsy of the brain, causing distension of the skull.

HYPOGONADISM Insufficient hormonal working of the reproductive glands.

HYPOPHYSIS Pituitary gland, a small hormonal gland situated at the base of the brain.

ILIUM The hip bone.

IMPETIGO Skin rash due to infection by staphylococci.

KAHLER'S DISEASE Malignant tumour, especially in the ribs, the skull, the shoulderblade and the spinal column.

KÜMMELL-VERNEUIL DISEASE Collapse of the vertebral body due to necrosis of the bone, often resulting from a minor but neglected fracture.

KYPHOSIS Bending of the spinal column resulting in a hunch-back.

LAMBDA The point on the cranium where the two parietals meet the occipital.

LIGAMENTUM INGUINALE Groin.

LIPOMA Benign tumour consisting of fat tissue.

LITTLE'S DISEASE A form of infantile spastic paralysis, probably due to cerebral haemorrhage at birth.

LUPUS ERYTHEMATOSUS A skin disease due to auto-antibodies in the circulation appearing as a result of the breakdown of normal immunization mechanism. Wholly distinct from lupus vulgaris which is a tuberculous skin infection.

LYMPHOGRANULOMATOSIS Hodgkin's disease. Malignant morbid growth of the lymph system.

MANDIBLE Lower jaw bone.

MANUBRIUM The upper part of the breast bone, or sternum.

MASTOID Osseous process behind the ear.

MENINGIOMA Cranial tumour starting from the meninges.

MESENCHYME Embryonic tissue from which the connective tissue develops.

METACARPUS Central part of hand.

METAPHYSIS Growth centre of the long bones, situated between the diaphysis and epiphysis.

METASTASIS Proliferation of a cancerous tumour.

METOPIC SUTURE A midline suture dividing the frontal bone in two halves. General in childhood, though it disappears later.

MITRAL STENOSIS Shrinking of the valve between the left auricle and left ventricle of the heart.

MUSCULUS DELTOIDES Muscle in the shoulder.

MUSCULUS QUADRICEPS FEMORIS A thigh muscle.

MUSCULUS SARTORIUS A thigh muscle.

MYELOMA Kahler's disease.

MYELOPLASTIC TUMOUR A tumour starting from the bone marrow.

MYOSITIS Inflammation of the muscle.

NASION The point where the bones of the nose and the forehead join each other.

NEOPLASM Literally 'new growth'; a term applied to any form of malignant tumour.

OBELION The point on the sagittal suture where the junction-line of the parietal arteries crosses.

OCCIPUT The bone at the back of the skull.

OLECRANON The elbow point of the ulna.

OPISTHION The middle point on the edge of the occipital hole.

OPISTHOTONOS Cramp position of the back and neck muscles causing the body to be arched backwards.

ORBITA Eye-sockets.

OSSA WORMIANA Small loose bones in the sutures of the skull.

OSTEITIS Bone inflammation.

OSTEOCHONDRITIS Bone and cartilage inflammation.

OSTEOMYELITIS Inflammation of the bone and bone marrow.

OSTEOPOROSIS Lessening of bone tissue with the bone going brittle.

PACHYMENINGITIS Thickening of the meninges after inflammation.

PARIETALS Two bones on the top of the cranium separated by the sagittal suture.

PARROT'S DISEASE Inflammation of the bone and cartilage (osteochondritis) through congenital syphilis.

PATELLA Kneecap.

PERIOSTITIS Infection of the periosteum.

PERIOSTEUM The membrane covering the bones.

PHAGOCYTOSIS The characteristic shown by some white blood-cells of consuming dead tissue cells, blood corpuscles and bacteria.

PHALANGES The bones of the finger.

PHIMOSIS Contraction of the foreskin, helped by circumcision.

PLASMOCYTOMA Swelling in the bone marrow due to plasme cells.

PLEURITIS Pleurisy: inflammation of the membrane lining the chest or covering the lungs.

POUTEAU-COLL'S FRACTURE A fracture of the forearm above the wrist joint with a staving in of the ulna process.

PROCESSUS SPINALIS VERTEBRAE Process of the vertebrae.

PREPUTIUM Prepuce, foreskin.

PROLAPSUS Bulging out of an internal organ.

PROXIMAL Lying closest to the centre: towards the rump. The opposite of distal.

PSOAS Relating to the loins.

PSORIASIS A skin disease with white scales.

PSEUDO-ARTHROSIS A false-joint caused by the failure of the bones to knit after a fracture.

PTERYGOID A wing-shaped skull-bone.

PUERPERAL SEPSIS General infection in women in labour.

PYORRHEA A pus-forming inflammation of the teeth.

RADIUS The outer bone of the forearm when the palm of the hand is uppermost.

RAPHE PERINEI Crutch.

RAYNAUD'S DISEASE Familial condition causing contraction of the arteries leading to necrosis of the fingers and toes.

RECKLINGHAUSEN'S (VON) DISEASE Ostitis fibrosa, a bone disease.

REITER'S DISEASE Arthritic infection caused by microbes in the intestines (enterococci).

SACRUM The part of the spinal column above the coccyx. It forms the posterior part of the pelvis.

SAGITTAL The cranial suture running in a front-back direction.

SARCOMA A form of malignant tumour.

SCAPULA Shoulder blade.

SCHEUERMANN'S DISEASE The appearance of a rounded back in adolescence.

SCLERODERMIA An incurable disease with a rapid course and unknown cause, causing the skin tissue to harden.

SELLA TURCICA The small hole at the base of the skull in which is the hypophysis.

SILICOSIS An inflammation of the lung due to working with stone.

SPHENOID Wedge-shaped skull bone.

SPINA BIFIDA Congenital malformation where the vertebral arches fail to close.

SPONDYLITIS RHIZOMELICA Inflammation of the spine at the level of the hip or shoulders.

SPONDYLOSIS Chronic inflammation of the spinal column with the forming of osseous processes.

STEATOPYGIA Enlargement of the buttocks, characteristic of Bushmen, especially female.

STERNUM The breast bone.

STEPHANION The point where the sutura coronalis crosses the sphenoid.

SULCUS Groove.

TABULA INTERNA AND EXTERNA Respectively the inner and outer layer of the cranial roof.

TALIPES EQUINOVARUS Clubfoot, so formed that the first part of the foot is used in walking.

TEMPORAL BONE The temple.

TIBIA The shinbone.

TRAUMA Injury due to physical violence.

THROMBOSIS The blocking of a blood vessel by a blood-clot.

TUBEROSITAS A rough process protruding from a bone.

ULNA The inner bone of the forearm, when the palm of the hand is uppermost.

URETHRA Urine canal.

VENA JUGULARIS The jugular vein of the neck.

M

| Era | | Geological Period | Duration (Millions of years) | Living Organisms |
|---|---|---|---|---|
| CENO-ZOIC | QUATER-NARY | Alluvium or Holocene | 0·01 | *Domestication. Beginning of civilization* |
| | | Diluvium or Pleistocene | 0·6 | *Palaeolithic man. Large mammals. Glacial periods.* |
| | TER-TIARY | Pliocene | 15 | *Man apes. Predecessors of present-day animal species.* |
| | | Miocene | 20 | *Mastodons and rhinoceros* |
| | | Oligocene | 10 | *Dog and Cat-like beasts of prey. Predecessors of horse, elephant, rhinoceros.* |
| | | Eocene | 25 | *Higher mammals. Disappearance of archaic forms.* |
| MESOZOIC or SECONDARY | | Cretaceous | 55 | *Decline of large reptiles, and specialization. Small mammals. Toothed birds.* |
| | | Jurassic | 25 | *Small mammals. Rise of birds and flying reptiles.* |
| | | Triassic | 50 | *Rise of dinosaurs.* |
| PALAEOZOIC or PRIMARY | | Permian | 25 | *Insects. First forms of reptiles.* |
| | | Carboniferous | 50 | *Development of amphibians.* |
| | | Devonian | 75 | *First animals on land (Amphibians) Land flora, Insects.* |
| | | Silurian | 100 | *First fishes, lung fishes, giant spiders, scorpions.* |
| | | Cambrian | 100 | *1,000 sorts of invertebrates. First known sea fauna: brachiopods, trilobites, corals, sponges, protozoa, molluscs, algae. No plants on land.* |
| ALGONKIUM | | Pre-Cambrian | 1500 | *Worms, radiolaria, bacteria.* |
| ARCHAICUM | | | | *No clear evidence of life.* |

TABLE I    Simplified diagram after Moodie, Post, Arambourg, and Senet.

| Period | Glacial era | Duration | Culture | Race | Fauna | Time-scale |
|---|---|---|---|---|---|---|
| HOLOCENE | | | Present-day Neolithic Mesolithic | Homo Sapiens | Present-day | |
| PLEISTOCENE | WÜRM—GLACIAL | −118,000 — 10,000 | Magdalenian Solutrean Aurignacian Mousterian Levalloisian | Homo Sapiens fossilis (type Cro-Magnon) | Elephas primigenius Rhinoceros tichorhinus | −10,000 |
| | | | Acheulian (final) | Homo Neanderthalensis | | −120,000 |
| | 3RD INTERGLACIAL PHASE | −180,000 — 118,000 | | | | |
| | RISS—GLACIAL | −230,000 — 180,000 | | | Elephas antiquus Elephas trogontherii Rhinoceros Merckii | −200,000 |
| | 2ND INTERGLACIAL PHASE | −420,000 — 230,000 | Acheulian Clactonian | | | |
| | MINDEL—GLACIAL | −480,000 — 420,000 | | Homo Heidelberg-ensis Pithecanthropus | Elephas cromerensis Rhinoceros etruscus | −450,000 |
| | 1ST INTERGLACIAL PHASE | −550,000 — 480,000 | Abbevillian (Chellian) | | | |
| | GÜNZ—GLACIAL | −600,000 — 550,000 | | | Elephas meridionalis | −600,000 |

TABLE II  Division of the Quaternary—taken freely from Camille Arambourg.

# Bibliography

1. ANDEREZ, V., *El craneo prehistórico de Santian* (Patronato de las Cuevas prehistóricas de la Provincia de Santander, 1954).
2. ANDEREZ, V., *Hacia el origen del Hombre* (Universidad Pontificia, Comillas, 1956, 390 pp.).
3. ANDREWS, R. Chapman, *Meet your ancestors* (London, 1944).
4. ÅSLANDER, A., Dental caries, the Bone-Meal Method and the cariogenic properties of sugar (*Tijdschr. Tandheelk*, 7, 1960, pp. 540–6).
5. AUFDERMAUER, M., Spondylite ankylosante I (*Documenta rheumatologica*, 2, 1954).
6. AUVRAY, LE DENTU, DELBET, *Nouveau traité de chirurgie. Maladies du crâne et de l'encéphale* 1909.

7. BAILLOUD, G. and MIEG DE BOOFSHEIM, P., *Les Civilisations Néolithiques de la France dans leur contexte européen* (Paris, 1955, 244 pp.).
8. BARRAL, C., Les hommes de la grotte Bianchi (*Bull. Mus. Anthrop. préhist. Monaco*, 3, 1956).
9. BARTELS, P., Tuberculose (Wirbelcaries) in der jüngeren Steinzeit (*Arch. f. Anthropologie*, 1907, pp. 243–55).
10. BARRIERE, Cl., *Les civilisations tardénoisiennes en Europe Occidentale* (Paris, 1956–7, 435 pp.).
11. BAUDOUIN, M., Les ossements de l'Allée Couverte de Vaudancourt (Oise) (*Mém. Soc. préhist. fr.*, IV, 1918–19).

12. BAUDOUIN, M., Une lésion inconnue de la rotule chez un préhistorique (*Paris Médical*, 96, 1935, pp. 470–3).
13. BAUDOUIN, M., La topographie des organes sexuels extérieurs féminins (*La Presse Médicale*, 1936, pp. 1907–8).
14. BEEX, G., Twee grafheuvels in Noord-Brabant (*Brabants Heem*, 1957).
15. BOULE, M. and VALLOIS, H. V., *Les hommes fossiles* (Paris, 1952, 583 pp.).
16 BRABANT, H., Trois cas d'agénésie de prémolaires au Néolithique (*Archs. Stomat., Liège*, 3, 1955).
17. BREUER, J., Furfooz et la Basse-Lesse (*Ardenne et Gaume*, XIII, 2, 1958).
18. BREUIL, H., *quatre cents siècles de l'Art pariétal* (Montignac).
19. BREUIL, H. and LANTHIER, R., *Les Hommes de la Pierre ancienne* (Paris, 1951, 335 pp.).
20. BREUIL, H., Découverte d'une grotte ornée paléolithique dans la province de Cacères (Nord-Ouest de l'Espagne) (*Bull. Soc. préhist. fr.*, 57, 1960, p. 155).
21. BRIAL, M., Anomalies morphologiques des dents (*Soc. d'études et de recherches préhistoriques et Institut pratique de préhistoire*, 9, 1960, pp. 139–58).
22. BROCA, P., Sur les trépanations du crâne et les amulettes crâniennes à l'époque néolithique (*Congrès international anthropol. et arch. préhist.*, Budapest, 1876, pp. 101–96).

23. BRUGHS, Th., *Lehrbuch der Inneren Medizin* (Berlin, 1942, 1534 pp.).

24. CARBALLO, J., *Prehistoria universal y especial de España* (Madrid, 1924).

25. CARBALLO, J., Esqueleto humano del periodo asturiense (*Invest. préhist.*, 2, 1960).

26. CARBALLO, J., Exploración de la gruta 'El Pendo' (Santander). (*Junta superior de excavaciones y antigüedades*, 1933).

27. CARTAILHAC, E. and BREUIL, H., *La caverne d'Altamira à Santillana près Santander* (Espagne) (Monaco, 1906).

28. CASTERET, N., *Dix ans sous terre* (Paris, 1951).

29. CHAUVET, S., *La médecine chez les peuples primitifs* (Paris, 1936).

30. CHIPAULT, A., *Chirurgie opératoire du système nerveux* (Paris, 1894).

31. CLAASSEN, A., Als mensen sterven (*Het Oude Land van Loon*, 1963, pp. 121–63).

32. DE LAET, S., L'Archéologie et ses problèmes (*Latomus*, 1950).

33. DE LAET, S. and GLASBERGEN, W., *De voorgeschiedenis der Lage Landen* (Brussel, 1959).

34. de LUMLEY, M., Les lésions osseuses de l'Homme de Castellar (A.-M.) (*Bull. Mus. Anthrop. préhist. Monaco*, 9, 1962, 191–205).

35. DE QUATREFAGES, A. and HAMY, E., *Crania Ethnica. Les crânes des races humaines* (Paris, 1882).

36. DEROBERT, L. and REICHLEN, H., *Les Momies* (Paris).

37. DESFOSSES, P., Paléontologie et Médecine (*Presse méd.*, 43, 1935, p. 2061).

38. DESSE, G. and GIOT, P., Lésions ostéoarticulaires de la nécropole gauloise de St Urnel-en-Plomeur (*Revue de Rhumatisme*, 11, 1952).

39. DESTEXHE-JAMOTTE, J., La sépulture néolithique d'Avennes (*Bull. Soc. r. Belg. Anthrop. préhist.*, 58, 1947, pp. 8–18).

40. D'HARCOURT, R., *La médecine dans l'ancien Pérou* (Paris, 1939).

41. DIRE, L., Les premières étapes et le chemin parcouru par la syphilis en France et en Europe (*Echo Médical*, 1959).

42. DÜGGELI, O. and TRENDELEN-BERG, F., La tuberculose de la colonne vertébrale (*Documenta rheumatologica*, 11, 1957).

43. DUJARDIN, B., *Propos sur la syphilis et son histoire* (Brussel, 1949).

44. DUPONT, E., Étude sur les cavernes des bords de la Lesse et de la Meuse, explorées jusqu'au mois d'octobre 1865 (*Bul. Acad. r. Belg.*, 20, p. 824).

45. DUPONT, E., Notice sur les fouilles scientifiques exécutées dans les cavernes de Furfooz (Prov. de Namur) (*Bull. Acad. r. Belg.*, 30, 1865, p. 244).

46. DUPONT, E., Étude sur l'ethnographie de l'homme de l'âge du renne (*Bull. Acad. r. Belg.*, 19, 1867).

47. ELAUT, L., *Het Medisch Denken* (Antwerpen, 1952).

48. FANCONI, G., Zur Pathomorphose der Poliomyelitis (*Triangel*, 7, 1956).

49. FRANCHET, L., Sur la dissolution des os et des dents dans les sépultures préhistoriques (*Revue anthrop.*, Paris, 1925, pp. 25–48).

50. FRANCHET, L. in MOREL, Ch., Le Tumulus N° X de Freyssinel (Causse de Sauveterre) (*Bull. Soc. préhist. fr.*, 4, 1934, pp. 177–94).

51. FRICKE, W., Untersuchungen an Leichenbränden der Gräberfelder von Prositz und Niederkaina (*Arbeits- und Forschungsberichte zur Sächsischen Bodendenkmalpflege*, 7, 1960, pp. 220–356).

52. GALLEZ, A. Le chêne à clous d'Herchies (Hainaut) (*Propagande et Tourisme*).

53. GASPARDY, G., Paläopathologische Untersuchungen an äneolithischen Skelettfunden in Ungarn (*Ethnographisch-Archäologischen Zeitschrift*, 1, 1961, pp. 1–32).

54. GASTAUT, H., Etiologie des épilepsies (*Encyclopédie médico-chirurgicale*, pp. 17008–30).

55. GEAY, P., Sur la découverte d'un squelette Aurignacien en Charente-Maritime (*Bull. Soc. préhist. fr.*, 54, 1957, pp. 193–7).

56. GEJVALL, N., Cremations (*Science in Archaeology*, 1963, pp. 379–90).

57. GERHARDT, K., Die Menschenreste aus den Gräbern von Gerlachsheim, Ilvesheim und Zeutern (*Badische Fundberichte*, 1958, pp. 161–72).
58. GLAZEMA, P., De Hunnebedden in Nederland (*Amersfoort*).
59. GIOT, P. and DESSE, G., Quelques documents sur les trépanations préhistoriques (*La Presse Médicale*, 72, 1950, pp. 1283–4).
60. GIOT, P. and COGNÉ, J., La nécropole de St. Urnel-en-Plomeur. Fouilles de 1946–50 (*Gallia*, 9, 1951).
61. GORDON, B., *Medecine throughout antiquity* (Philadelphia, 1949, 830 pp.).
62. GOSLINGS, J. and HIJMANS, W., La signification du phénomène L.E. dans la polyarthrite chronique évolutive (*J. belge Méd. phys. Rhum.*, 13, 1958, pp. 113–25).
63. GOURY, G., *Origine et Evolution de l'Homme*, Tome I, Epoque paléolithique (Paris, 1948).
64. GRAHMANN, R., *Préhistoire de l'Humanité* (Paris, 1955).
65. GRIMM, H., Der gegenwärtige Stand der Leichenbranduntersuchungen (*Ausgrabungen und Funde*, 6, 1961, pp. 299–306).
66. GUIARD, E., *La trépanation crânienne chez les Néolithiques et chez les primitifs modernes* (Paris, 1930).

67. HAECK, L. and JANSSENS, P., La grotte du Mont Falise à Antheit. (*Bull. Soc. r. Belg. Anthrop. Préhist.*, 74, 1963).
68. HEER, O., Die Pflanzen der Pfahlbauten (*Neuj. Bl. naturf. Ges. Schaffhausen*, 1865).
69. HEMMETER, E., Médecins et Thérapeuthique dans l'Egypte Ancienne (*Ciba Revue*, 19, 1941).
70. HOFSCHLAEGER, R., Von den Krankheiten der vorgeschichtlicher Menschen (*Ciba Zschr.*, 1939, 6 pp. 2319–24).
71. HOUZÉ, E., Les Cimetières francs de Belgique (*Bull. Soc. Anthrop. Bruxelles*, 1891–2).
72. HOUZÉ, E., Les Néolithiques de la Province de Namur. (*Congrès d'Archéologie et d'Histoire*, Dinant, 1903).

73. IVANICEK, F., *Staroslovenska Necropola U Ptuju* (Ljubljana, 1951).

74. JÄGER, K., *Beiträge zur frühzeitlichen Chirurgie* (Wiesbaden, 1907).
75. JANSSENS, P., Medical Views on Prehistoric Representations of Human Hands (*Medical History*, 1, 1957, pp. 318–22).
76. JANSSENS, P., Figures anatomiques dans l'art préhistorique (*Bull. Soc. r. Belg. Anthrop. Préhist.*, 69, pp. 84–90).
77. JANSSENS, P. and ECHEGARAY, J., *Memoria des las excavaciones de la Cueva del Juyo (1955–56)* (Patronato de las cuevas prehistor. de la Prov. de Santander (España), 1958).
78. JANSSENS, P., La transición del arpon magdalenense al arpon acilense (*Investigaciones prehistóricas II*, Santander, 1960, pp. 164–73).
79. JANSSENS, P., Onderzoek der Crematieresten uit tumulus IV, p. 25 in BEEX, H. Onderzoek van Grafheuvels te Weelde (*Toxandria*, 30, 1958, pp. 3–29=Archaeologia Belgica, nr. 47).
80. JANSSENS, P., De crematieresten uit de grafheuvels op de Roosen te Neerpelt (1959) in ROOSENS, H. en BEEX, G. Onderzoek van het urnenveld op de 'Roosen' te Neerpelt in 1959 (*Limburg*, 39, 1960, pp. 59–142=Archaelogia Belgica nr. 48).
81. JANSSENS, P., Het squelette néolithique d'Avennes – Sa perforation sternale (*Bull. Soc. r. Belg. Anthrop. Préhist.*, 71, 1960, pp. 43–6).
82. JANSSENS, P., Onderzoek van de crematieresten uit het urnenveld 'De Roosen' te Neerpelt in 1960, pp. 28–35, in ROOSENS, H. en BEEX, G. De opgravingen in het Urnenveld 'De Roosen' te Neerpelt in 1960 (*Het Oude Land van Loon*, 16, 1961=Archaeologia Belgica nr. 58).
83. JANSSENS, P., Onderzoek der crematieresten uit een tumulus te Grobbendonk, in MERTENS, J. Gallo-romeins graf uit Grobbendonk (*Scrinium Lovaniense*, 1961, pp. 136–49=Archaeologia Belgica nr. 53).
84. JANSSENS, P., Onderzoek der crematieresten uit twee urnen, gevonden te Eksel, pp. 165–6, in DE LAET, S. Opgravingen en vondsten in de Limburgse Kempen (*Limburg*, 40, 1961=Archaeologia Belgica nr. 55).

85. JANSSENS, P., Examen des restes incinérés découverts à Tontelange, pp. 25-6 in BONENFANT, P. Sépultures trévires à Tontelange (*Archaeologia Belgica* nr. 57).

86. JANSSENS, P., De crematieresten pp. 150-60 in ROOSENS, H. en BEEX, G. De Opgravingen van het Urnenveld 'De Roosen' te Neerpelt in 1961 (*Het Oude Land van Loon*, 17, 1962, pp. 145-73 = Archaeologia Belgica nr. 65).

87. JANSSENS, P. and ROOSENS H., Lijkverbranding en lijkbegraving op het Merovingisch grafveld te Grobbendonk (*Helinium*, 3, 1963, pp. 265-72).

88. JANSSENS, P., Onderzoek der beenderresten van de opgravingen te Destelbergen (crematieresten) in DE LAET, S., VAN DOORSELAER, A. en SPITAELS, P. Oudheidkundige opgravingen en vondsten in Oost-Vlaanderen (*Kultureel Jaarboek voor de Provincie Oost-Vlaanderen*, 2, 1963).

89. JANSSENS, P., Anthropologisch onderzoek der beenderresten van de inhumatie- en krematiegraven, in VANVINCKENROYE, W. *Gallo-romeinse Grafvondsten uit Tongeren* (Tongeren, Provinciaal Gallo-romeins Museum, 1963).

90. JANSSENS, P., De crematieresten uit het urnenveld te Grote-Brogel in ROOSENS, H., BEEX, G. en BONENFANT, P. Een urnenveld te Grote-Brogel (*Limburg* 42, 1963, pp. 261-300 = Archaeologia Belgica nr. 67).

91. JANSSENS, P., Trépanations préhistoriques (*Bull. Soc. r. Belg. Anthrop. Préhist.*, 70, 1959, pp. 69-81).

92. JANSSENS, P., Het eerste arbeidsongeval in België (*Informatiebulletin Hoofddienst V.G.V.*, 8-9, 1960, pp. 82-3).

93. JANSSENS, P., La Race de Furfooz: son âge, sa pathologie (*Bull. Soc. r. Belg. Anthrop. Préhist.*, 73, 1963, pp. 45-55).

94. JANSSENS, P., Anthropologisch onderzoek van de crematieresten uit de grafheuvels te Hamont, in ROOSENS, H. en BEEX, G. Bronstijdgrafheuvels op de Haarterheide te Hamont (*Archaeologia Belgica* nr. 81, 1965).

95. JESSERER, H. and KIRCHMAYR, W., L'ostéoporose d'involution présénile et sénile (*Documenta Rheumatologica*, 8, 1956).

96. JUDE, P., Sépultures et rites funéraires chez les Paleo-pléolithiques (*Société études recherches Préhistor. et Institut pratique de Préhist.*, 1960, pp. 177-83).

97. KHARADLY, M., Oud-Egyptische Geneeskunde (*Ciba-symposium*, 3, 1956, pp. 66-72).

98. KLAATSCH, A., *Die Anfänge der Kunst und Religion der Urmenschheit* (Leipzig, 1913).

99. KREFT, S., Untersuchungen an jungsteinzeitlichen Kinderskeletten (*Arbeits- und Forschungsberichte zur Sächsischen Bodendenkmalpflege*, 5, 1956, pp. 23-56).

100. KROGMAN, W., The medical and surgical practises of Pre- and Protohistoric Man (*Ciba-Symposium*, 2, 1940, pp. 444-52).

101. KROGMAN, W., The pathologies of Prehistoric and Protohistoric Man (*Ciba-Symposium*, 2, 1940, pp. 432-43).

102. KÜHN, H., *Das Erwachen der Menschheit* (Frankfurt, 1954).

103. KÜHN, H., *De kunst van het Oude Europa* (Utrecht/Antwerpen, 1959).

104. KUZELL, W. and GAUDIN, G-P., La goutte (*Documenta Rheumatologica*, 10, 1956).

105. LE BARON, J., *Lésions osseuses de l'homme préhistorique en France et en Algérie* (Paris, 1881, 131 pp.).

106. LE BARON, J., Sur les lésions osseuses préhistoriques (*Soc. Anthrop. Paris*, 4, 1881, pp. 597-8).

107. LE DOUBLE, A., *La médecine et la chirurgie dans les temps préhistoriques* (Paris, 1911).

108. LEROY-GOURHAN, A., *Les Hommes de la Préhistoire. Les Chasseurs.* (Paris, 1955).

109. LE ROUZIC, Z., *Carnac* (Vannes, 1939, 138 pp.).

110. LINDNER, K., *La chasse préhistorique* (Paris, 1957, 480 pp.).

111. LUQUET, G., Les Vénus Paléolithiques (*Journal de Psychologie*, 31, 1934, pp. 429-60).

112. LUQUET, G., La magie dans l'art paléolithique (*Journal de Psychologie*, 28, 1931).

113. MACCURDY, G., Human skeletal remains from the Highlands of Peru. (*Am. J. Phys. Anthrop.*, 6, 1923, pp. 217–329).

114. MACLAREN, J., *My crowded solitude* (London, 1954).

115. MANOUVRIER, L., Le T-sincipital. Curieuse mutilation crânienne néolithique (*Bull. Soc. Anthrop. Paris*, 4e ser., 6, 1895, p. 357).

116. MARIËN, M., *Oud-België* (Antwerpen, 1952).

117. MARTIN, R., *Lehrbuch der Anthropologie in systematischer Darstellung* (Jena, 1957, 2762 pp.).

118. MARTY, J., LAGARDE, C., PERROT, J. and ESQUIREL, F., Monstrueux anévrysme extériorisé de l'Aorte (*La Presse Médicale*, 68, 1960, pp. 1232–4).

119. MASKA, Ch., OBERMAYER, H. and BREUIL, H., La statuette de Mammouth de Predmost (*L'Anthropologie*, 23, 1912, p. 285).

120. MAUDUIT, J., *40.000 ans d'art moderne* (Paris, 1954).

121. MEROC, L., Nouvelle découverte de peintures dans la grotte de Tibiran, Comm. de Tibiran-Jaunac (Hautes-Pyrénées) (*Bull. Soc. préhist. fr.*, 53, 1956, p. 46).

122. MICHEZ, J., Les conditions physiques étiologiques influençant l'apparition de la P.C.E. (*J. belge. Méd. phys. Rhum.*, 13, 1958, pp. 87–102).

123. MILLER, J., Some diseases of Ancient Man (*Ann. Med. Hist./1*, 1929, pp. 394–402).

124. MOODIE, R., *Paleopathology*, an introduction to the study of ancient evidences of disease (Urbana, Illinois, 1923, 567 pp.).

125. MOORE, R., *Man, Time and Fossils* (Chicago).

126. MOREL, Ch., Le tumulus n° X du Freyssinel (Causse de Sauveterre) (*Bull. Soc. préhist. fr.*, 4, 1934, pp. 177–94).

127. MOREL, Ch. (Son), *La médecine et la chirurgie osseuses aux temps préhistoriques dans la région des Grands Causses* (Paris, 1951).

128. MOUQUIN, M. and BARDIN, P., Les artérites aigues (*Encyclopédie médico-chirurgicale*, 8, 1936, p. 11315).

129. MØLLER-CHRISTENSEN, V. Ein jungsteinzeitliches Skelett mit mehreren geheilten Schädelverletzungen von Grosz Upahl, Kr. Güstrow (*Ausgrabungen und Funde*, 5, 1960, pp. 171–4).

130. NELISSEN, A., Liaison entre la préhistoire et le folklore: silex taillé retrouvé dans une habitation actuelle. (*Bull. Soc. préhist. fr.*, 51, 1954, p. 32).

131. NOUGIER, L. and ROBERT, R., Un 'foyer tribal' du Magdalénien Pyrénéen (*La Nature*, 3253, 1956).

132. NOUGIER, L. and ROBERT, R., *Mas-d'Azil* (Toulouse).

133. NOUGIER, L., *Géographie humaine préhistorique* (Paris, 1959, 325 pp.).

134. PALES, L. and WERNERT, P., Une mandibule pathologique de grand bovidé du Loess d'Achenheim (Bas-Rhin) (*Annales de Paléontologie*, 39, 1953).

135. PALES, L., *Paléopathologie et pathologie comparative* (Paris, 1930, 352 pp.).

136. PLANCQUAERT, R., Une palafitte et une nécropole de palafitteurs à Overpelt (Limbourg) (*Bull. Soc. Anthrop. Bruxelles*, 39, 1924, pp. 203–7).

137. PERICOT GARCIA, L. *L'Espagne avant la conquête romaine* (Paris, 1957, p. 299).

138. PERICOT GARCIA, L., *La cueva del Parpallo* (Madrid, 1942, 351 pp.).

139. POPP, H., Krankheiten und Chirurgie des Urmenschen (*Med. Welt*, 13, 1939, pp. 127–9).

140. POST, J., *De Wieg der Mensheid* (Amsterdam, 308 pp.).

141. RADMILLI, A., Steatopygia in the prehistoric female figure (*Rassegna Medica*, 12, 1955, p. 140).

142. RAHIR, E., Les habitats et les sépultures préhistoriques de la Belgique (*Bull. Soc. Anthrop. Bruxelles*, 60, 1925).

143. RAHIR, E., Découvertes archéologiques faites à Furfooz de 1900 à 1902 (*Bull. Soc. Anthrop. Bruxelles*, 1914).

144. RAYMOND, P., Les maladies de nos ancêtres à l'âge de la pierre (*Aesculape*, 2, 1912, pp. 121–3).
145. RIAD, N., *La médecine au temps des Pharaons* (Paris, 1955, 319 pp.).
146. RIVET, L., Pathologie et chirurgie préhistoriques (*Presse med.*, 53, 1945, p. 402).
147. ROBERT, R., Deux œuvres d'art inédites de la grotte de la Vache (Ariège) (*Bull. Soc. préhist. fr.*, 3–4, 1951).
148. ROGER, J., Quelques remarques sur la dynamique des populations et la paléontologie (*Bull. Mus. Hist. nat.*, Paris, 27, 1955, pp. 153–9).
149. ROUILLON, A., *Lésions osseuses préhistoriques de la Vendée* (Angers, 1923).
150. ROUSSY, G., LEROUX, R. and OBERLING, C., *Précis d'anatomie pathologique* (Paris, 1929).
151. RUST, A., *Die Alt- und Mittelsteinzeitlichen Funde von Stellmoor* (Neumünster, 1943).
152. RUST, A., Die jüngere Altsteinzeit im Jungpaläolithicum (*Historia Mundi*, 1, 1952).
153. RUTOT, A., *Un essai de reconstitution plastique des races humaines primitives* (Brussel, 1919).

154. SACCASYN-DELLA SANTA, E., *Les figures humaines du Paléolithique supérieur Eurasiatique* (Antwerpen, 1947).
155. SCHMERLING, P., Description des ossements, à l'état pathologique, provenant des cavernes de la province de Liège (*Bull. Soc. géol. fr.*, 7, 1835, pp. 51–61).
156. SCHMIDT, B., Gräber mit trepanierten Schädeln aus frühgeschichtlicher (*Jschr. mitteldt. Vorgesch*, 47, 1963, pp. 383–7).
157. SCHULTZ, A., Notes on diseases and healed fractures of wild apes and their bearing on the antiquity of pathological conditions in man (*Bull. med. Hist.*, 7, 1939, pp. 571–2).
158. SENET, A., *L'Homme à la recherche de ses ancêtres* (Paris, 1954, 343 pp.).
159. SEUNTJENS, H., Sur la portée d'une interprétation totémiste des figurations humaines paléolithiques (*Bull. Soc. préhist. fr.*, 53, 1957, pp. 609–17).

160. SHETELIG, H. and FALK, H., *Scandinavian Archaeology* (Oxford, 1937).
161. SIGERIST, H., *A History of Medecine* (Oxford, 1951).
162. STECHER, R., Heredity of Rheumatoid Arthritis (*Journal belge Méd. Rhum.*, 13, 1958, pp. 103–11).
163. STEKLÁ, M., Sépultures du peuple à céramiques spiralée et pointillée (*Archeologické Rozhledy*, 2, 1956, pp. 697–723).
164. SUDHOF, K., *Geschichte der Medizin* (Berlin, 1922).

165. TAXIL, J., *Traité de l'épilepsie, maladie vulgairement appelée la goutette aux petits enfants* (Lyon, 1603).
166. TECOZ, H., Les maladies de l'homme préhistorique (*Praxis*, 33, 1944, p. 174).
167. TWIESSELMANN, F., *Les représentations de l'homme et des animaux quaternaires découvertes en Belgique* (Brussel, 1951).

168. VALLOIS, H., La durée de la vie chez l'homme fossile (*C.R. Acad. sc.*, 204, 1937, pp. 60–2).
169. VALLOIS, H., La durée de la vie chez l'homme fossile (*Anthrop.*, 47, 1937, pp. 499–532).
170. VALLOIS, H., Les maladies de l'homme préhistorique (*Revue scientifique*, 72, 1934, pp. 666–78).
171 VAN DEN BROECK, J., *De Dageraad der Mensheid* (Utrecht, 1950, 219 pp.).
172. VAULTIER, R., La Médecine populaire en Lorraine (*La Presse Médicale*, 1960).
173. VERBRUGGE, A., Les figurations de mains humaines dans l'art paléolithique (*Bull. Soc. r. Belg. Anthrop. Préhist.*, 64, 1953, pp. 185–91).
174. VERHEYLEWEGHEN, J., Le Paléolithique final de culture périgordienne du gisement préhistorique de Lommel (*Bull. Soc. r. Belg. Anthrop. Préhist.*, 57, 1956).
175. VERHEYLEWEGHEN, J., Un dépôt funéraire de crâne néolithique à Spiennes (Hainaut) (*Helinium*, 2, 1962, pp. 194–214).

176. VERNEAU, R., La prétendue parenté des Négroides européens et des Boschimans (*L'Anthropologie*, 1925, p. 235).

177. VERWORN, BONNET and STEINMANN, *Der Diluviale Menschenbefund von Obercassel, bei Bonn* (Wiesbaden, 1919).

178. VLÊCK, E., Contribution à l'analyse anthropologique des sépultures incinérées (*Archeologické Rozhledy*, 2, 1956, pp. 724–7).

179. VON KÖNIGSWALD, G., *Les premiers hommes sur la terre* (Paris, 1956).

180. WAKEFIELD, E. and DELLINGER, S., Diseases of prehistoric Americans of South Central United States (*Ciba Symposia*, 2, 1940).

181. WALKER, A., Problems in post-traumatic epilepsy (*Archives of Neurology and Psychiatry*, 59, 1948, pp. 254–8).

182. WEISS, H., *Casos peruanos prehistóricos de cauterizaciones. T-sincipital de Manouvrier* (Lima, 1955).

183. WELLS, C., A study of cremation (*Antiquity*, 34, 1960).

184. WELLS, C., A case of lumbar osteochondritis from the bronze age (*Journal of Bone and Joint Surgery*, 43, B, 1961, p. 575).

185. WELLS, C., A new approach to ancient disease (*Discovery*, 1961, pp. 526–31).

186. WELLS, C. and MAXWELL, B., Alkaptonuria in an Egyptian mummy (*The British Journal of Radiology*, 35, 1962, pp. 679–82).

187. WELLS, C. in SMEDLEYN and OWLES, E., Some Suffolk Kilns: IV Saxon Kilns in Cox Lane (*The proceedings of the Suffolk Institute of Archaeology*, 29, 1963).

188. WELLS, C., Ancient Egyptian pathology (*The Journal of Laryngology and Otology*, 77, 1963, pp. 261–5).

189. WELLS, C., Cortical grooves on the Tibia (*Man*, 137, 1963, pp. 112–14).

190. WELLS, C., The study of ancient disease (*Surgo*, 32, 1964).

191. WELLS, C., *Bones, Bodies and Disease* (Ancient Peoples and Places, London, 1964).

192. WENDT, H., *A la recherche d'Adam* (Paris, 1954).

193. WERTHEIMER, P., AVET, J., LEVY, A. and JENOT, J., Les lacunes osseuses de la voûte crânienne (*La Presse Médicale*, 68, 1956, pp. 1556–9).

194. WOUTERS, A., Een kaakfragment van Cervus giganteus met ingeschoten gravettespits (*Ber. Rijksdienst. oudheidk. Bodemonderz*, 8, 7–10, 1958).

# Index

*Figures in italics indicate most important references.*

Abbevillian phase, *10*
Abscess, 91
Acheulean, *10*
Achondroplasia, 8, 49, 53
Acromegaly, 50
Actinomycosis, 87, 90, 115
Adipose, 56
Aegean peoples, 12
Aesculapius, temple of, 5
Afalou-Bou-Rhommel, 61
Ahrensburgian, *11*, 12, 40
Ainu, 88, 133
Alexandria, 37, 68
Algérie, 29
Algonkian rocks, 88
Alkaptonuria (ochronosis), 53
Almerian culture, 12
Altai, 8
Altamira, 11, 38, 39, 120, 143, 148
Anatomia, *37*
Andalusian art, 39
Andes, 8
Aneurism, aortic, 48
Anglo-Saxon, 48
Antheit, 31
Anthropology, 4, 17, 18, 22, 47
Apatosaurus, 68, 72
Aphelops, 115
Appendicitis, 119
Archaeopteryx, 74
Archaicum, 1, 2
Ariège, 123
Art (prehistoric), *38*
Arteriosclerosis, 35, 117, 123
Arteritis, 123
Arthritis, 65, *75*
Asia Minor, 4
Assuan, 100
Atlantic phase (climate), 12
Aurignac(ian), *10*, 11, 35, 38, 39, 40, 42,
   44, 56, 57, 59, 83, 91, 93, 120, 123

Australopithecines, 26
Avennes, *48*
Avitaminosis, *64*
Azilian, *11*, 44

'Bandkeramik', 18, 22, 59
*Bâton de commandement*, 13, *58*, 147
Bédeilhac, 120
Bechterew's disease, 79, 80
Beni-Hassan, grave of, 65
Bernissart, 72
Berula Kodo, 122
Besnier-Boeck's disease, 89
Blastomycosis, 116
Brassempouy, 56
Broken Hill (Rhodesia), 26, 73, 95
Bronze Age, 12, 36, 53, 99
Buerger's disease, 124
Bushmen, 56
Byzantine periods, 94

Cabrerets, 120
Calabria, 13
Callus, 25, 27, 30, 68
Cambrian, 17
Canine, 92
Cannibal, 18, 64
Cape Circeo, 88, 132
Carabelli's cusp, 92, 109
Carboniferous, 2
Carcinoma, 14, 69
Caries, *93*
Castillo, cave of, 41, 120
Chapelle-aux-Saints, La., 91
Chatelperron, *10*
Cheddar culture, *11*
Chimu period (Peru), 8
Choukoutien, cave of, 26, 132
Ceratopsia, 85
Clactonian, *10*
Clinker, 23

Colombres (Spain), 137
Congenital syphilis, 47
Covalanas, cave of (Spain), 40
Coxalgia, 98
Craniolacuna, 126
Cremation, *22*, *23*
Creswell culture, *11*
Cretaceous, 67, 85, 90
Cribra cranii, 52
Cribra orbitalia, 52
Cro-Magnon, 9, 26, 27, 31, 59, 60, 62,
    83, 86, 88, 142

Deformation (cranial), *96*
Dermoid cyst, 127
Devonian, 2, 88
Dimetrodon, 25
Dinosaurus, 74, 82, 85, 98
Dordogne, 123
Dwarf, 49
Dyaks, 133
Dysostosis, 126

Ecuador, 7
Egypt, 12, 31, 37, 49, 51, 84, 86, 109, 118
Embalming, 13, 37, 71
Enamel, 92
Enzypteridae, 17
Eocene, 77
Epidaurus, 5
Epilepsy, 134, 136
Equihen (France), 35
Erasistratus, 37
Eskimo, 64
Ewing's tumour, 127
Exloo-Odoorn, 12
Exostosis, 67

*Fieberkrämpfe*, 134
Fistula, vesico-vaginal, 118
Folklore, 45
Fracture, 25, 28, 32, 33
Framboesia (Yaws), 14, 50, 104
Friedlander's disease, 123
Furfooz, 18, 20, 21, 32, 67, 73, 91, 144

Gallstones, 118, 119
Gargas, cave of, 8, 120, 122
Gizeh, 68
Gloucestershire, 48
Goundou, 50, *115*
Gout, *89*, 119
Gramat (France), 61
Gravette (point), *10*, 35
Grimaldi, caves of, 18
Grote Brogel (Limburg), 23

Gummata, 106
Günz, glacial era, 9
Gynecomastia, 53

Haematuria, 118
Hallux valgus, 89
Hamburgian culture, *11*
Hand, *120*
Harris's lines, 15
Hastière (Belgium), 21, *132*
Heidelberg, 9, 100
Herchies (Belgium), 46
Herodotus, 13
Herophilos, 37
Hoëdic, 61
*Höhlengicht*, 86
Hogeloon (burial site), 22
Hippocrates, 4
Holocene, 9, 11
l'Homme-Mort, cave of, 28, 98
Hottentot, 121
Hunchback, 101
Hydatid cyst, 127
Hydrocephalus, 48
Hypercementosis, 91
Hyracodon, 91

Iguanodons, 72
Iron Age, 9, 12, 23, 29, 33, 45, 62, 149

Jivaro Indians, 7

Kerckering, node of, 55
Kersaint-Plabennec, 135
Khasi, of Assam, 7
Kidney, 118
Köln-Lindenthal, 59
Krapina (Yugoslavia), 84
Kümmell-Verneuil's disease, 99

Lascaux, cave of, 40, 41, 143, 144, 148
Laussel, 56, 57
Leishmaniasis, 116
Limeuil (Dordogne), 140
Leprosy, *114*
Lespugue, 56
Levalloiso-Mousterian, *10*
Lizières, 137
Lombrives, 29
Lozère, 49, 99
Luiksgestel, 22
Lymnocyon potens, 66
Lyngby, antler-axe, 26, 141

Magdalenian, *11*, 35, 39, 41, 59, 94, 148
Maglemosian, *12*

Magritte, Trou (Belgium), 42, 56
Malaria, *115*
Mandam Indians, 121
*Marchets*, 21
Mas d'Azil, cave of, 87
Maya, 51
Megalithic, 33
Meningeal artery, 35
Meningeoma, 69
Meningocoele, 126
Merovingian culture, 7, 22
Meryhippus campestris, 90
Mesolithic, 9, 35, 51, 59, 61, 62, 94,
    95, 125, 137, 141
Mesopotamia, 9
Metastases, 69, 70
Metopic sutures, 47
Meudon, dolmen of, 68
Microlithic, 11
Mindel glacial era, 9
Miocene, 90, 115
Mixnitz, 68, 87
Mochica period (Peru), 8
Montfort, cave of, 35
Montardit, 61
Mosasaur, 67, 85, 90
Mousterian, 41, 87
Mouthe, La, cave of, 38
Mummification, 7, 89
Mutilation, 121
Myeloma, 89
Myeloplastic growth, 69
Myositis ossificans, 85

Natufian period (Palestine), 62
Neanderthal, 9, 26, 31, 57, 60, 61, 62, 63,
    73, 83, 84, 86, 87, 88, 90, 91, 93, 95,
    142
Neerpelt, 22, 83
Neolithic, 4, 7, 18, 22, 23, 27, 31, 32, 33,
    35, 36, 40, 46, 48, 50, 53, 59, 60, 64, 65,
    67, 68, 71, 73, 82, 83, 87, 91, 92, 94,
    95, 96, 98, 99, 100, 112, 117, 125, 131,
    132, 134, 135, 136, 137, 138, 141, 149
Neoplasm, 67
Niaux, 41
Nigeria, 8
Nigua, *116*
Nogent-les-Vierges, 125

Obercassel, 26
Obourg, Man from, 4, 31
Ochronosis, 53
Ofnet, cave of, 26, 64, 140, 141
Olecranon, 48
Opisthotonos, 74

Osteitis, 32, 50
Osteochondritis, 53
Osteoma, 68
Osteomalacia, 66
Osteomyelitis, 14, 66, 72, 73
Osteoporosis, 50, 52, 69
Osteosarcoma, 67
Overpelt (Belgium), 132

Paget's disease, 23, 41, 50
Palaeolithic, 9, 18, 26, 35, 40, 42, 51, 56,
    61, 64, 65, 71, 73, 91, 92, 94, 95, 100,
    102, 140, 146, 148, 149
'Palaeophytopathology', 2
Palaeozoic, 1, 88
Pair-non-Pair, 68
Pansinusitis, 73
Parasitosis, 2, 18
Parpallo, cave of, 40
Parrot's disease, 107, 109
Parry fracture, 18
Pasiega, La, 41
Patallacta, 129
Paulus of Aegina, 111
Paviland, caves of, 68
Pech de l'Aze, 132
Pekin man, 26, 140
Pelvis, 118
Pendo, El, 58
Penis, 42
Perigordian, *10*, 42, 56
Periost(itis), 67, 72
Permian, 25
Pes equinovarus, 111
Phimosis, 42
Phytosaur, 25
Pick's disease, 126
Pindal, cave of, 40, 41, 123, 137
Pithecanthropus, 9, 25, 31
Plague, *114*, 118
Platecarpus, 67
Platycnemia, 109
Pleistocene, 9, 64, 68, 79, 88, 93, 98
Pleurisy, 118
Pliocene, 79
Pneumonia, 118
Polar regions, 8
Poliomyelitis, *111*
Pompeii, 7
Population density, 58
Pott's Disease, 99
Pre-Cambrian, 89
Pre-Columbian, 12, 28, 48, 49, 51, 55,
    64, 68, 72, 73, 83, 84, 91, 92, 93, 94,
    100, 104, 106, 129
Predmost, 141

Preputium, 42
Primary, 86
Pterodactylus, 74
Ptolemaic period, 94
Ptuju, 65
Pygmies, 121

Quaternary, *9*, 93, 85

Raynaud's disease, *124*
Recklinghausen's disease, von, 49, 50, 52
Reiter's disease, 76
Rheumatic arthritis, 14
Riss glacial era, 9
Roda, 108
*Rondelle,* 125, 130, 131, 132

Santian, cave of, 142
Sarcoma, 69
Sauropodes, 85
Schleswig-Holstein, 46
Schistosoma, 118
Sclaigneaux (Belgium), 98
Scoliosis, 99
Scurvy, *64*
Scythian (Proto), burials, 8
Secondary, 13, 77, 82
Siberia, 8
*Signe tectiforme,* 41
Silicosis, 4
Sinanthropus, 9, 64, 132
Sincipital-T, *138*
Sinusitis, 91
Sleeping-sickness, *115*
Smallpox, *114*
Smilodon californius, 85
Solutrean, *11*, 39, 40, 41, 59, 83, 107
Sordes (Landes), 27
Spiennes, 4, 7, 46
Spina bifida, 48, 49, 126
Spondylarthrosis, 53
Spondylitis, 76, 79, 81, 82
Steatopygia, 56

Steinheim, 132
Stone weapons, *35*
St Urnel-en-Plomeur, 60
Sweating sickness, 16
Syphilis, 14, 50, 51, *103*, 123, 126

Talipes equinovarus, *49*
Tardenoisian, *12*, 35
Tayacian, *10*
Tertiary, 82, 91
Téviec (Brittany), 35, 61
Thessaly, 4
Tiberan, cave of, 121
Tibio-tarsal, 66
Tjongerian culture, *11*
Toledo, 13
Tollund Man, 7, 20
Tongeren, 32
Tradition, *45*
Trauma, *25*
Trepanation, *125*
Trois-Frères, cave of, 120
Tuberculosis, *98*
Tuc d'Audoubert, cave of, 43, 88
Tumour, *67*, 127

Uricaemia, *89*
Uta, *116*
Utrecht, 46

Vagina, 42
Vaudancourt, 21
'Venus' figurines, 8, *56*, 132
Verruga Peruana, *116*
Vertebrates, 1
Vesuvius, 1
Vienne, cave of, 55

Washakia, 66
Weimar, 132
Würm glacial era, 9

Yaws, 14, 104